THE LOW COST CARRIER WORLDWIDE

The Low Cost Carrier Worldwide

Edited by
SVEN GROSS
Hochschule Harz – University of Applied Sciences, Germany
and
MICHAEL LÜCK
AUT University, New Zealand

ASHGATE

Published by
Ashgate Publishing Limited
Wey Court East
Union Road
Farnham
Surrey, GU9 7PT
England

Ashgate Publishing Company
110 Cherry Street
Suite 3-1
Burlington, VT 05401-3818
USA

www.ashgate.com

British Library Cataloguing in Publication Data
The low cost carrier worldwide.
1. Airlines--Rates. 2. Airlines--Cost of operation.
3. Aeronautics, Commercial--Law and legislation.
I. Gross, Sven, 1969- II. Lück, Michael, 1966-
387.7'12-dc23

The Library of Congress has cataloged the printed edition as follows:
Gross, Sven, 1969-
The low cost carrier worldwide / by Sven Gross and Michael Lück.
p. cm.
Includes bibliographical references and index.
ISBN 978-1-4094-3268-5 (hbk) -- ISBN 978-1-4094-3269-2 (ebk) --
ISBN 978-1-4724-0022-2 (epub) 1. Airlines--Rates. 2. Airlines--Cost of operation.
3. Airlines--Economic aspects. 4. Aeronautics, Commercial--Law and legislation.
I. Lück, Michael, 1966- II. Title.
HE9783.G76 2013
387.7'42--dc23

2013000257

ISBN 978 1 4094 3268 5 (hbk)
ISBN 978 1 4094 3269 2 (ebk – PDF)
ISBN 978 1 4724 0022 2 (ebk – ePUB)

MIX
Paper from responsible sources
FSC® C013985
FSC www.fsc.org

Printed in the United Kingdom by Henry Ling Limited,
at the Dorset Press, Dorchester, DT1 1HD

Contents

List of Figures

List of Tables

Notes on Contributors

Yazmin Aguilar López, Universidad Anáhuac, Mexico
Mtra. Yazmín Aguilar López is currently Professor at the tourism school at Anáhuac University in Mexico City. She completed a Tourism Business Administration Bachelor's degree, has a diploma in Human Resources and a Master's degree in Education. She works with SABRE de México (GDS) as instructor.

Email: yaguilar@anahuac.mx

Borislav Bjelicic, DVB Bank and University of Mannheim, Germany
Borislav is Head of Corporate Communications at DVB Bank SE, a leading specialist in the international transport finance business. Before he joined DVB Bank SE, he had worked for Deutsche Lufthansa German Airlines from 1990 to 1998. Borislav studied Business Administration, majoring in Logistics and Transportation Management, followed by graduation as a Doctor of Economics in 1990. He has been a lecturer on Transportation Management at the University of Mannheim since 1995. In 2001 he was appointed Honorary Professor by the University of Mannheim. He received the Rhenania Logistics Award in 1985, the Academic Award of the Stinnes Foundation in 1990, and the Award of the Prechel Foundation in 1991. He has authored more than 60 articles and contributions to books on transportation topics.

Email: Borislav.Bjelicic@dvbbank.com

Roland Conrady, University of Applied Sciences, Worms, Germany
Roland Conrady has been Professor in the Department of Tourism and Transportation at Worms University of Applied Sciences since 2002. His research and teaching focus is on air travel, e-business and tourism. Roland has also been director of the world's largest tourism congress, the ITB Berlin Congress, since 2004, and is on the supervisory board of several tourism companies. He is a member of the board of the German Society of Tourism Research (DGT) and author of several books (including publications such as *Luftverkehr*). He was previously in charge of the Electronic Business course and Professor of General and Transport Business Administration at Heilbronn University of Applied Sciences. After gaining his doctorate at Cologne University in 1990, he worked for Deutsche Lufthansa in several executive positions until 1998.

Email: conrady@fh-worms.de

Montserrat Flores Zozoaga, Universidad Anáhuac, Mexico
Montserrat Flores Zozoaga is currently studying the last year of the Bachelors Degree in Gastronomy at Universidad Anáhuac México Norte.

Sven Gross, University of Applied Sciences, Harz, Germany
Sven Gross is Professor of Transport Carrier Management at the Harz University of Applied Sciences and Deputy Director of the Institute for Tourism Research in Wernigerode, Germany. Since August 2008 he has been co-director of the Tourism Department in the Competence Center of Information and Communication Technologies, Tourism and Services at Harz University, and in 2007 he joined the New Zealand Tourism Research Institute (NZTRI), based in Auckland, New Zealand. Sven studied Tourism Geography and Economics at the University of Trier, and Spatial Planning at the University of Dortmund. He was a management and community consultant for several years, personal assistant to the mayor of Bad Duerkheim and Research Associate at the Dresden University of Technology, where he was awarded his PhD. His main research areas include tourism and transport (e.g. airline management, cruise ship management, car rental management), business travel management and tourism market research. He is author of more than sixty publications on these subjects, including nine books.

Email: sgross@hs-harz.de

Uwe Hermann, Tshwane University of Technology, South Africa
Uwe P. Hermann is a Lecturer in Tourism Management in the Department of Tourism Management at the Tshwane University of Technology (TUT) in Pretoria, South Africa. He holds an M-Tech degree in Tourism and Hospitality Management from TUT and is currently pursuing a PhD at North West University, Potchefstroom. His research focus is on sustainable tourism and events, heritage tourism and commercial aviation.

Email: HermannUP@tut.ac.za

Markus Landvogt, Kempten University of Applied Sciences
In 2012, Markus was appointed Professor for Information Management in Tourism at the Kempten University of Applied Sciences in Bavaria. Between 2006 and 2012, Markus worked as a Senior Research Analyst in the Tourism Strategy Group at the Ministry of Economic Development in New Zealand, previously the Ministry of Tourism. There, he was involved in managing the publicly funded tourism statistics system of New Zealand. Prior to moving to New Zealand he was for nine years Professor of Information Technology in Leisure, Tourism and Transport at Stralsund University of Applied Sciences in Germany. Besides his main subject he also taught courses on aviation industry and revenue management in tourism. In addition, Markus has gained a wide range of practical work experience in the tourism sector, having run his own tour operator business in

Germany, an IT consultancy for tour operators and an upmarket boutique B&B in New Zealand. His main research interests are in quantitative methods in tourism, in particular in seasonality in tourism, revenue management and tourism statistics. His was awarded a PhD in applied mathematics at Mainz University in Germany.

Email: markus.landvogt@fh-kempten.de

Gustavo Lipovich, Universidad de Buenos Aires, Argentina
Gustavo Lipovich is a specialist in air transport geography. He is a researcher in the Transport and Territory Programme and Lecturer in Transport Geography at the Universidad de Buenos Aires (Argentina). He has a Masters degree in Urban Economics (UTDT-Argentina) and a Doctorate in Geography (UBA-Argentina). He is the author of several articles published in different journals and congresses on the subject of the relationship between air transport and economic development. Gustavo was one of the founders of the Ibero-American Network of Air Transport Research (RIDITA), and was President from 2006 to 2009. He has worked as a consultant, as an assessor with Aerolíneas Argentinas, and is currently a member of the Aeropuertos Argentina 2000 Board of Directors, representing the National State shares participation.

Email: glipovich@yahoo.com.ar

Gui Lohmann, Southern Cross University, Australia
Gui Lohmann is a Lecturer in the School of Tourism and Hospitality Management, Southern Cross University, Australia. He earned his PhD at Victoria University of Wellington, New Zealand, and has taught and undertaken research in several universities around the world, including the University of Waikato, New Zealand, Universidade de São Paulo, Brazil, and the University of Hawaii at Manoa, USA. He is the author of several books and peer-reviewed journal articles about the relationship between transport and tourism. Gui has worked as a consultant for the Brazilian Ministry of Tourism, the World Tourism Organization, the United Nations Environmental Programme and the Abu Dhabi Tourism Authority. He is the Founding Executive Director of ABRATUR, the International Academy for the Development of Tourism Research in Brazil.

Email: Lohman.gui@gmail.com

Michael Lück, Auckland University of Technology, New Zealand
Michael Lück earned his PhD at the University of Otago, New Zealand, and currently holds the position of Associate Professor in the School of Hospitality and Tourism, AUT University in Auckland, New Zealand. He is Head of the Tourism & Events department, and an Associate Director of the New Zealand Tourism Research Institute (NZTRI). In 2012, Michael joined the Institute for Tourism Research at the Harz University of Applied Sciences in Wernigerode, Germany.

Michael's primary research interests are in the wider area of coastal and marine tourism, with a focus on marine wildlife tourism, interpretation and education. He is also interested in ecotourism, sustainable tourism, the impacts of tourism, aviation, and gay tourism. Michael has developed a keen interest in innovative and alternative teaching and assessment methods. He has published in international academic journals and contributed to various books. Michael is co-author of a tourism text, the overall editor or co-editor of seven books and the *Encyclopedia of Tourism and Recreation in Marine Environments* (CABI); the Founding Editor of the academic journal *Tourism in Marine Environments*; and Associate Editor of the *Journal of Ecotourism*.

Email: mlueck@aut.ac.nz

Brooke Porter, Auckland University of Technology, New Zealand
Brooke Porter is a PhD candidate at Auckland University of Technology in New Zealand. Her primary research interest is in developing access points into sustainable tourism markets for communities in less developed nations and understanding drivers behind resource dependencies. Brooke's other areas of interest include ecotourism, interpretation in wildlife tourism, voluntourism, fisheries management, service learning and community outreach.

Email: emailbrookey@gmail.com

Marius Potgieter, Tshwane University of Technology, South Africa
Marius Potgieter is a Senior Lecturer in the Department of Tourism Management at the Tshwane University of Technology (TUT) and his fields of interest are aviation, tourism marketing, and international destinations. Marius worked in industry for 11 years at South African Airways, Translux, and Quicksilver Coach Lines and continued his studies and based his doctoral thesis on marketing information systems. Marius regularly presents papers at international academic conferences and also provides study leadership to postgraduate students.

Email: PotgieterM@tut.ac.za

Alexander Oliver Scherer Leibold, Universidad Anáhuac, Mexico
Mtro. Alexander Oliver Scherer Leibold is currently the coordinator of the Bachelors Degree of Gastronomy at Universidad Anáhuac. He completed a Master´s degree in marketing and publicity and is currently working on his PhD in innovation and social responsibility. He has published articles in magazines about gastronomy and service, and has written a book entitled *The Customer and Service Quality*.

Email: ascherer@anahuac.mx

Alexander Schröder, DB Netz AG, Germany
Alexander Schröder is working for DB Netz AG, the rail infrastructure company of Deutsche Bahn AG in Germany. Previously he worked as a lecturer at the Harz University of Applied Sciences in Wernigerode, Germany, and as a research associate with the Chair of Tourism Economics and Management at the Dresden University of Technology. He also worked as a product manager for the Mediterranean area of a package tour operator. He is the author of several publications on the subjects of safety and security in tourism, business tourism and low cost carriers.

Email: alexander.schroeder@deutschebahn.com

Tarun Shukla, *Mint*, *Hindustan Times*, India
Tarun Shukla currently works as a National Writer at *Mint* (*The Wall Street Journal's* partner newspaper in India). In his eight years of being a journalist, Tarun has covered the aviation sector in depth (besides defence, energy and media) and won various in-house and international accolades. His keen nose for investigative stories got him the prestigious The Society of Publishers in Asia (SOPA) 2011 Award for excellence in investigative reporting. Tarun's passion for travelling and curiosity about the aviation sector led him to author a book on the evolution of IndiGo, India's largest low-cost airline, which parallels the rise of India's economy.

Email: tarunsmail@gmail.com

Semisi Taumoepeau, AIS St Helens, New Zealand
Semisi Taumoepeau is Head of the Department of Arts, Tourism and Hospitality Management at AIS St Helens in Auckland (2003–present date). Over the years Semisi has worked in the Pacific as Director of Tourism in Tonga, as a CEO of Royal Tongan Airlines and as Chairman of the South Pacific Tourism Organisation based in Fiji. He holds a BSc in Chemistry (Auckland), an MSc in Tourism Planning and Development (Surrey) and a Doctorate from the University of the Sunshine Coast (Queensland), where he wrote his thesis on the 'Economic Sustainability of South Pacific Airlines'. He is a consultant and a researcher to NZAID and has undertaken several projects in the Pacific Islands. Most of his recent publications were on Pacific tourism and aviation in the Asia Pacific region. He recently published a book on Pacific airlines entitled *South Pacific Aviation.*

Email: semisit@ais.ac.nz

Timothy M. Vowles, University of Northern Colorado and Colorado State University, USA
Timothy M. Vowles is a Visiting Assistant Professor in the Geography Department at the University of Northern Colorado and the Anthropology Department at

Colorado State University. Tim earned his PhD in Geography from the University of Denver, where his dissertation focused on the impact of Southwest Airlines on pricing and service patterns. His Master's degree, also in Geography, centred on creating a model to predict the loss of commuter air services in the United States and won the Transportation Research Forum's best graduate student paper. He has worked on numerous transport research projects, including a project funded by the National Center for Intermodal Transportation, 'Assessing Intermodal Transportation Planning at State Departments of Transportation', and 'An Evaluation Of Air–Rail Passenger Intermodal Access at United States' Airports'. His research interests include the role of low-cost carriers such as Ryanair and Southwest Airlines, changing air service patterns and the role global air alliances play in service and pricing patterns around the world. Another area of research interest is the role of airports as intermodal gateways to metropolitan areas and regions, along with the emergence and classification of secondary and niche airports within an air transportation system. Tim has published research articles focusing on air transport issues in numerous journals, including the *Transportation Law Journal*, the *Journal of Air Transportation Management,* the *Journal of Transport Geography*, and the *Professional Geographer.*

Email: Timothy.Vowles@unco.edu

Andreas Wald, European Business School, Paris, France
Andreas Wald is Dean of Faculty and Research, Professor of Management and Strategy at the European Business School, Paris, and a Visiting Professor at the Strascheg Institute for Innovation and Entrepreneurship of the EBS University, Germany. He teaches undergraduate and graduate courses in Strategic Management and Organization. Andreas has published numerous articles in peer-reviewed journals, books and edited volumes. He is the Editor-in-Chief of the *International Journal of Aviation Management*. His research is focused on organizational networks, temporary organizations, innovation, and the aviation industry.

Email: andreaswald@ebs-paris.com

Preface

Although low cost carriers (LCCs) can trace their roots back to more than 40 years ago, they are a relatively young phenomenon. It is only recently that LCCs experienced an increase in academic interest in the form of books, journal articles, and research reports.

This edited volume aims to provide an overview of the development of LCCs around the globe. In an attempt to achieve this lofty goal a number of aviation experts have contributed chapters, discussing regional aviation profiles, the background and development of LCCs in the respective markets, the development of the low cost market, recent trends in these markets and best practice examples.

As with all edited books, this volume came to life with the support of many individuals and organizations. Of course, we are indebted to all contributing authors who freely gave their time and expertise to work on specific regional sections. At times it was a challenge for us to coordinate such a diverse group of authors with a range of backgrounds and procedures. However, we are lucky to have found specialists from all continents who have written sections about their respective home markets. These are: Yazmin Aguilar López, Borislav Bjelicic, Roland Conrady, Montserrat Flores Zozoaga, Uwe Hermann, Markus Landvogt, Gustavo Lipovich, Gui Lohmann, Brooke Porter, Marius Potgieter, Alexander O. Scherer Leibold, Alexander Schröder, Tarun Shukla, Semisi Taumoepeau, Timothy M. Vowles, and Andreas Wald.

We would like to sincerely thank Ashgate Publishing, and in particular Guy Loft. From the outset Guy supported this book enthusiastically and was very lenient and understanding when we were unable to meet self-set deadlines. We would also like to thank the reviewers who provided feedback on our initial proposal and gave us important ideas, enabling us to ensure the quality of this book.

A very special thank-you goes to Louisa Klemmer, Anne Menzel and Brooke Porter for their administrative support, editorial help and translations. We could not have done it without you!

We are equally lucky to have working environments that are supportive of such ventures. Sven would like to thank Axel Dreyer and his colleagues in the Competence Center of Tourism at the University of Applied Science Harz. Micha would like to thank Linda O'Neill (Head, School of Hospitality and Tourism), Nigel Hemmington (Dean, Faculty of Culture and Society and Pro-Vice Chancellor, International), and his colleagues in the Department of Tourism and Events (all at AUT University, Auckland).

Last, but most certainly not least, we would like to thank our families, who often get the short end of the stick when we are sitting in the office working on books such as this, but amazingly never complain about it. Without your loving support we could not have done it!

Sven Gross, Wernigerode
Michael Lück, Auckland

List of Abbreviations

AACO	Arab Air Carriers Organization
AAI	Airports Authority of India
ABRATUR	Academia Internacional para o Desenvolvimento da Pesquisa em Turismo no Brasil [The International Academy for the Development of Tourism Research in Brazil]
ACAC	Arab Civil Aviation Commission
ACARS	Aircraft Communications Addressing and Reporting System
ACSA	Airports Company South Africa
AFCAC	African Civil Aviation Commission
AFTK	Available Freight Tonne Kilometres
AICM	Aeropuerto Internacional Benito Juárez de la Ciudad de México [Mexico City's International Airport]
AIG	American International Group
AMAIT	Administradora Mexiquense del Aeropuerto Internacional de Toluca [Mexican Administration of the International Airport of Toluca]
AMU	Arab Maghreb Union
ANAC	Agencia Nacional de Aviaçáo [Brazilian Civil Aviation National Agency]
APEC	Asia-Pacific Economic Cooperation
ASA	Air Service Agreement
ASA	Aeropuertos y Servicios Auxiliares [Airport and Auxiliary Services]
ASEAN	Association of Southeast Asian Nations
ASK	Available Seat Kilometres
ASUR	Aeroportos del Sureste de México [Mexican Southeast Airport Group]
ATC	Air Traffic Control
ATI	Air Transport Intelligence
ATR	Aerei da Trasporto Regionale/Avions de transport régional [Regional Tranport Aircraft]
BA	British Airways
BAG	Banjul Accord Group [Sierra Leone Aviation Safety Oversight Organisation]
BiH	Bosnia and Herzegovinia
BOAC	British Overseas Airways Corporation
BWI	Baltimore-Washington International Airport
CAA	Civil Aviation Authority
CAAC	Civil Aviation Administration of China

CAB	[US] Civil Aeronautics Board
CEMAC	Commission de la Communauté Economique et Monetaire de l'Afrique Centrale [Economic Community of Central African States]
CIS	Commonwealth of Independent States
Comair	Commercial Air Services
COMESA	Common Market for Eastern and Southern Africa
CRJ	Canadair Regional Jet (c.f. Bombardier CRJ)
CWC	Carriers Within Carriers
DAC	Departamento da Avação Civil [Brazilian Department of Civil Aviation]
DGAC	Dirección General de Aeronáutica Civil [Mexican Directorate General of Aeronautics]
DGCA	Directorate General of Civil Aviation [India]
DGT	Deutsche Gesellschaft für Tourismuswissenschaft e.V. [German Society of Tourism Research]
DLR	Deutsches Zentrum für Luft- und Raumfahrt [German Aerospace Centre]
DOT	[US] Department Of Transportation
ECAA	European Common Aviation Area
EAA	East African Airways Corporation
EAC	East African Community
EBIT	Earnings Before Interest and Taxes
ECAA	European Common Aviation Area
ECOWAS	Economic Community of West African States
ELFAA	European Low Fares Airline Association
ETS	[European Union] Emissions Trading Scheme
EU	European Union
EU-8	The eight Member States that joined the EU in 2004.
FAA	[US] Federal Aviation Administration
FFP	Frequent Flyer Programme
FLF	Freight Load Factor
FSA/FSC	Full Service Airline/Carrier
FSNC	Full Service Network Carrier
FTK	Freight Tonne Kilometres
GACM	Grupo Aeroportuario de la Ciudad de México [Mexico City Airport Group]
GAP	Grupo Aeroportuario del Pacífico [Pacific Airport Group]
GCC	Gulf Cooperation Council
GDP	Gross Domestic Product
GDS	Global Distribution System
GOL	Gol Linhas Aéreas
IATA	International Air Transport Association
ICAO	International Civil Aviation Organization
ICT	Information and Communication Technologies

IFALPA	International Federation of Air Line Pilots' Associations
INSEE	National Institute of Statistics and Economic Studies (France)
IPO	Initial Public Offering
JAL	Japan Airlines
JCA	Joint Competitive Authority
KMIA	Kruger Mpumalanga International Airport
Lanseria	Lanseria International Airport
LCA/LCC	Low Cost Airline/Low Cost Carrier
M&A	Mergers and acquisitions
MDG	Millennium Development Goal
MENA	Middle East and North Africa
NEPAD	New Partnership for Africa's Development
NTA	National Transportation Agency, Canada
NZAID	New Zealand Aid Programme
NZTRI	New Zealand Tourism Research Institute
OMA	Grupo Aeroportuario del Centro-Norte [formerly Operadora Mexicana de Aeropuertos]
ONS	Office for National Statistics, UK
PLF	Passenger Load Factor
PKP	Passenger Kilometres Performed
PSA	Pacific Southwest Airlines
RIDITA	Ibero-American Network of Air Transport Research
rpk	revenue passenger kilometres
SAA	South African Airways
SADC	Southern African Development Community
SA Express	South African Express Airways
SANTACO	South African National Taxi Council
SATCC	Southern African Transport and Communications Commission
SCT	Secretaría de Comunicaciones y Transportes [Mexican Ministry of Communication and Transport]
SIA	Singapore Airlines
SOPA	The Society of Publishers in Asia
TAM	Transportes Aéreos del Mercosur S.A.
TCH	Transport Clearing House
THAI	Thai Airways
TUT	Tshwane University of Technology
UAE.	United Arab Emirates
WTTC	World Travel and Tourism Council
UEMOA	L'Union Economique et Monétaire Ouest Africaine/West African Economic and Monetary Union
UNECA	United Nations Economic Commission for Africa
UNWTO	United Nations World Tourism Organization
VARIG	Viação Aérea Rio Grandense S.A.
VASP	Viação Aérea São Paulo

VFR	Visiting Friends and Relatives
WTO	World Tourism Organization
WTTC	World Travel and Tourism Council
YD	Yamoussoukro Decision

PART I
Introduction

Chapter 1
The Low Cost Carrier – A Worldwide Phenomenon?!

Sven Gross, Michael Lück and Alexander Schröder

Introduction

An overview of the low cost carriers (LCCs) operating within the worldwide air traffic market becomes outdated (almost) as quickly as editions of a daily newspaper over the course of the day of publication. Nonetheless, from both an academic and a practitioner's perspective it is not only interesting but also important to gain an insight into LCC developmental trends, possible continental differences and the identification of typical elements of the low cost business model.

The following analysis is based on a comprehensive review of publicly available information regarding the worldwide operation of LCCs and information gathered from industry publications.[1] Furthermore, individual LCCs were contacted via telephone and/or email for clarification purposes or to gain additional information.

Classification of airlines according to the low cost model was also based on industry publications (see Note 1). It was determined whether the airlines were currently operating and whether they applied typical characteristics of low cost business models. Typical regional airlines acting as feeder airlines within local markets (such as Egypt Air Express and American Eagle) and charter airlines (such as Air Holland, Condor and TUIfly) were not included in the analysis. However, some airlines have developed into a hybrid model, for example Air Berlin serves typical low cost routes, yet also offers long haul flights as well as typical tourist routes, usually served by charter airlines. Furthermore, while Aer Lingus operates as a LCC within the European market, it more closely resembles a full service carrier (FSC) on long haul flights; such hybrid carriers were included for the purpose of this analysis. Subsequently, 109 LCCs[2] currently active in the worldwide market were identified.

---

1 In particular, *World Airlines 2010* (Flight International 2010); *Global LCC Outlook Report* (Harbison and McDermott 2009); *Low Cost Carrier Profiles* (ATI 2008); The Airline Industry Guide 2009/10 and 2010/11 (*Airline Business* 2010, 2011); *The Low Cost Monitor* (DLR 2012); and the European Low Fares Airline Association's Current Airline Membership list (ELFAA 2012).

2 An overview of low cost carriers included in the analysis can be found in Appendix 1.1.

Table 1.1 Numbers of low cost carriers per continent (2011)

Continent	Currently active	Active for a limited time	Never started/Startup
Africa	8	2	1
Asia	35	7	1
Europe	37	34	8
Middle East	5	2	3
North America	11	19	0
Oceania	5	6	2
South America	8	8	1
Total	*109*	*78*	*16*

Additionally, 94 LCCs were identified which were either active for a limited time (78), or announced their intentions but never actually carried out flight operations (16). Table 1.1 illustrates the LCCs included in the current analysis grouped according to continents. The majority of currently active LCCs are operating in Europe and Asia.

Development of the Low Cost Carrier

As is the case with many developments within air transportation, the LCC concept originated in the USA. Until the mid-1970s, the regulatory provisions of the aviation market were based on the Civil Aviation Act of 1938, which codified strict regulations regarding market access and price controls. The strict regulations resulted in exorbitant prices, distorted pricing structures to the detriment of long haul flights, and brought about an intensive and costly competition over quality (Aberle 2003). In the early 1960s the former US Civil Aeronautics Board (CAB) first approved the variation of promotional fares based on time of day, length of stay, or type of travel. However, the market remained rather rigid and it was not until the early 1970s that greater flexibility was achieved. This was due mainly to the introduction of new types of aircraft (such as the Boeing B747, which doubled the available seating capacity in comparison to the previously used aircraft); and the increasing activity of charter airlines, which entered the market with extremely low fares. In the 1970s the CAB deregulated the American aviation market. Due in particular to pressure exerted by charter airlines wishing to sell their fares without the requirement to sell packages including additional travel services, the American aviation market was deregulated by the CAB in the 1970s.

While in 1975 the CAB eased regulations regarding market access, in 1978 an important milestone was achieved with the ratification of the Airline Deregulation Act (Pompl 2007). Important transportation policy objectives were to increase competitiveness through liberalization of market access; to achieve greater flexibility of pricing structures by eliminating requirements; and to bring about greater adaptability of the airlines to current market conditions (Aberle 2003).

Based on the economic developments, the resulting increase in demand for air travel, and the deregulation of the market, there was an increase in competitiveness within the aviation market. Former charter airlines and new start-ups entered the market as LCCs, often offering dumping prices, and engaged in direct competition with the traditional (FSCs).

It is rather difficult to pinpoint the origins of LCCs since opinions in the literature differ (see also Figure 2.1 in Chapter 2 of this volume). For example, Pacific Southwest Airlines (PSA) founded in 1949 as a Californian intrastate carrier, was transformed into an LCC in the early 1960s; unfortunately, due to poor strategic decision making and investments PSA was bankrupt by the mid-1970s (Bailey, Graham and Kaplan 1985).

Southwest Airlines, founded in 1967 and since developed into one of the largest airlines in the world, is also regarded as one of the pioneers in the low cost sector. Southwest started operations on 18 June 1971 and decided against the conventional network and hub system, instead opting for easily implementable point-to-point services (Bjelicic 2004, Knorr 2007).

Following Southwest Airlines, further LCCs were founded in the USA such as American TransAir (1972) and Midwest Express (1984), while already established companies entered the low cost market with subsidiary enterprises.[3] In 1972, the British Laker Airways intended to enter the low cost segment with a 'Skytrain' operating between London and New York, but operations did not begin until 1977 when Laker received traffic rights for the United States. Thereafter further LCCs such as Braniff (1979), Virgin Atlantic (1984) and People Express (1983), began operations in the low priced air traffic segment between the two continents. Although the concept was not economically successful – with the exception of Virgin Atlantic all other airlines exited the market – it significantly changed the market (Pompl 2007).

Growth vs Decline

The respective deregulation processes had an accelerating effect, so that during the 1990s and 2000s LCCs gradually established themselves on all continents. As shown in Table 1.2, there were two decisive time periods for the development of LCCs: the early to mid-1990s in Canada, Europe and Oceania; and the 2000s in Asia, Eastern Europe and the Middle East. Thus, it is of no great surprise that most LCC start-ups were launched within these time periods (see Figure 1.1). As has previously been mentioned, 94 LCCs are no longer active or were founded

3 These airlines have not been included in the analysis, with the exception of Southwest Airlines, because they are still successfully operating.

Table 1.2 The beginning of deregulation and low cost developments by regions

Region (country)	Year low cost operations began (First Airline/Pioneer)	Implementation of deregulation
USA	1971 (Southwest Airlines)	1978
EU	1986 (Ryanair)*	1986
Australia	1990 (Compass Airlines, withdrew 1991)	1990
New Zealand	1994 (Kiwi Travel International Airlines, withdrew 1996)	1984
Canada	1996 (WestJet)	1996
Japan	1998 (Skymark Airlines)	1998
Malaysia	2001 (Air Asia)	2001
Brazil	2001 (GOL)	1998
South Africa	2001 (Kulula)	1999
EU Expansion	2002 (Skyeurope, withdrew 2009)	2004
Gulf States	2003 (Air Arabia)	2003
India	2003 (Air Deccan, withdrew 2008, JetKonnect)	2003
Thailand	2004 (Nok Air, Thai AirAsia)	2003
Singapore	2004 (Tiger Airways, Valuair)	2001
China	2005 (Spring Airlines)	Ongoing

Note: * The deregulation of air transport between Britain and Ireland in 1986 brought reductions in fares of over 50 per cent and a doubling of passenger numbers,in contrast to high fare increases and market stagnation before deregulation (Barrett 1997). Ryanair obtained 1986 permission from the regulatory authorities to challenge the British Airways and Aer Lingus high fare duopoly on the Dublin–London route. Services were launched with two (46-seater) turboprop BAE748 aircraft. The first flights operated in May from Dublin to London Luton. After three years of rapid growth in aircraft, routes and intense price competition with Aer Lingus and British Airways, Ryanair accumulated £20 million in losses and went through a substantial restructuring in 1990: the Ryan family invested a further £20 million in the company, and copying the Southwest Airlines low fares model the airline was relaunched under new management as Europe's first low cost airline (see http://www.ryanair.com/en/about).

Source: Francis et al. 2006 with amendments based on information garnered from individual airlines.

but never fully began operations.[4] From Figure 1.1 it is clear that most of the companies exited the market within the period 2004–2005 and during the global economic crisis (2008–2009). The most LCCs were recorded before the global economic crisis. Interestingly, there were also many new start-ups during the

4 An overview of the 94 LCCs can be found in Appendix 1.2.

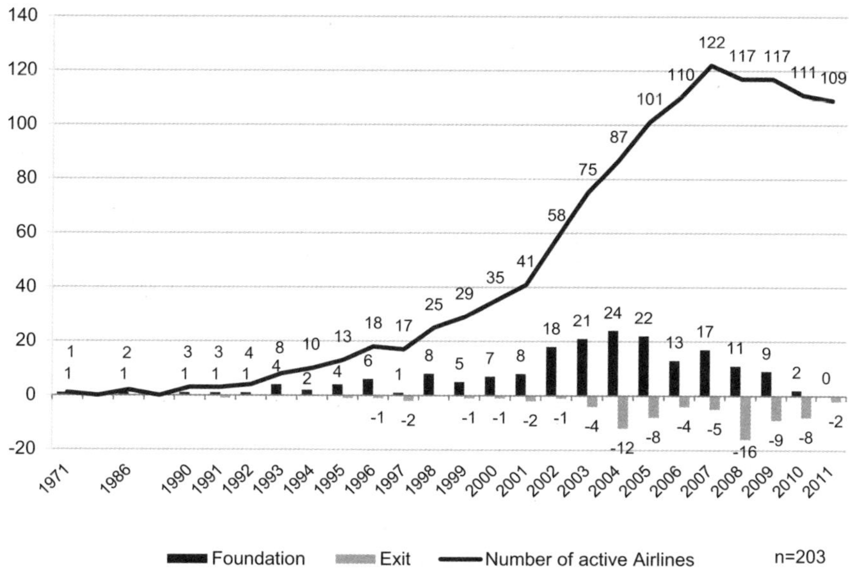

Figure 1.1 Number of market entries and exits of low cost carriers

middle of the decade, which can be attributed to the worldwide expansion of the low cost business model previously described.

Within this context the reasons cited for exiting the market are also of interest. As can be expected, financial problems are of the utmost importance: 43 per cent of the companies which exited the market did so due to bankruptcy. The suspension of operations among LCCs due to economic deficits/red figures/high losses takes second place (18 per cent). This measure was often taken in connection with the

Table 1.3 Number of market exits and their reasons by continent (1991–2011)

	Bankruptcy	Discontinuation	M&A	Start Up/ Never started	Sum
Africa	1	0	1	1	3
Asia	4	0	3	1	8
Europe	18	4	12	8	42
Middle East	1	1	0	3	5
North America	8	10	1	0	19
Oceania	3	2	1	2	8
South America	5	0	3	1	9
Sum	40 (43%)	17 (18%)	21 (22%)	16 (17%)	94

establishment of a low cost subsidiary by a FSC (often referred to as carrier within carrier), as well as a restructuring of the airline, i.e. a modification of the business model to charter or to a FSC. Other reasons are mergers and acquisitions (M&A) through another airline (22 per cent).

Additionally, 16 airlines were identified which never started flight operations and have since disappeared from the market (never started) or are listed as start-ups (mid-2011) and are preparing flight operations (see Table 1.3 on the previous page).

A detailed examination of the companies exiting the aviation market reveals that the greatest fluctuation took place within the European market; almost half of the companies exiting the LCC market can be found in this region. Furthermore, approximately two-thirds of the airlines that never commenced operations can be found in the combined markets of the Europe and the Middle East.

Analysis of Corporate Organization: Newcomer Versus Subsidiary Airlines

Some LCCs founded further airlines as subsidiary companies and have been fairly successful. For example, the Virgin Group operates as an FSC (Virgin Atlantic) and low cost subsidiaries in Oceania (Virgin Australia) and America (Virgin America).

Over the last few years some FSCs have also founded their own subsidiaries or had holdings/shares in LCCs (see Table 1.4). This measure was taken mostly as a reaction to LCCs operating within their own market (e.g. Qantas launched Jetstar and Singapore Airlines founded Tiger Airways).

The Lufthansa Group for example, includes the LCC Germanwings and, since 2008, Lufthansa has also had stakes in the American low cost airline JetBlue (see Table 1.5).

Table 1.4 Selected low cost carrier subsidiaries and holdings/shares of other low cost carriers

Low cost mother	Low cost subsidiary (operation started)
Air Arabia	Air Arabia Maroc (2009), Air Arabia Egypt (2010)
Air Asia	Thai AirAsia (2004), Indonesia AirAsia (2005), AirAsia X (2007, together with Virgin Group)
easyJet	easyJet Switzerland (1999)
Ryanair	Viva Aerobus (2007, Ryanmex 49 %)
Virgin Group LCC subsidiaries	Virgin Express (1996, ceased operations in 2007) Virgin Blue (2000)*, Pacific Blue (2004)*, Polynesian Blue (2005)*, Virgin America (2007), Virgin Australia (2009)*
Wizz Air	Wizz Air Bulgaria (2005), Wizz Air Ukraine (2008)

Note: * All Virgin subsidiaries in Australia, New Zealand and Oceania have been combined under the Virgin Australia brand since 2011.

Table 1.5 Selection of active full service carrier subsidiaries, either as start-ups or as holdings/shares of other airlines (sometimes via subsidiaries)

FSC mother	Low cost subsidiary (operation started)
Air France	Transavia France (2007)
Alitalia	Air One (2010)
Asiana	Air Busan (2008)
Emirates Group	Flydubai (2009)
Iberia	Vueling Airlines (2004)
Jet Airways	JetKonnect (2003) JetLite (2007)
KLM/Basiq Air	Transavia Airlines (2005)
LAN	Aires (2009)
Philippine Airlines	Airphil Express (1996)
Singapore Airlines	Tiger Airways (2004) Tiger Airways Australia (2007)
Turkish Airlines	Anadolu Jet (2008)
Air India	Air India Express (2005)
All Nippon Airways	Air Japan (2001), Skynet Asia Airways (2002)
British Airways/Comair	kulula.com (2001)
Garuda	Citilink (2001)
JAL	JAL Express (1998)
Kingfisher	Kingfisher Red (2008)
Korean Air	Jin Air (2008)
Lufthansa	bmibaby (2002) Germanwings (2002) JetBlue, 19% share since 2008 (2000)
Qantas	Jetstar Airways (2004) Jetstar Asia Airways (2004) Jetstar Pacific (2007)
Thai Airways	Nok Air (2004)

However, a number of low cost subsidiary start-ups by FSCs were not successful. In fact, almost all of the major American airlines failed with such attempts (see Table 1.6 on the next page).

Analysis of Classic Elements of the Low Cost Business Model

Due to the multiplicity of airlines which operate under the label 'low cost', there is not one distinct low cost business model. Within the academic/scientific debate however, a number of elements typical to LCCs have been identified. These

Table 1.6 Selection of bankrupt and/or discontinued subsidiaries of full service carriers

FSC mother	Low cost subsidiary	Operation started	Operation withdrawn	Years at market
Air Canada	Tango	2001	2004	3
	Zip	2002	2004	2
Air New Zealand	Freedom Air	1995	2008	13
	CityJet	1999	1999	0.5
All Nippon Airways	Air Next	2005	2010	6
British Airways	Go	1998	2002	4
	DBA	2003	2008	5
Continental Airways	Continental Lite	1993	1995	2
Cyprus Airways	Hellas Jet	2002	2005	3
Delta Airlines	Delta Express	1996	2003	7
	Song Air	2003	2006	3
Finnair	Fly Nordic	2001	2009	8
Hapag Lloyd	HLX	2002	2007	5
KLM	Buzz	2000	2003	3
LAN Airlines	LAN Express	1998	2004	6
LOT Airlines	Centralwings	2004	2009	5
Mexicana	Mexicanalink	2009	2010	1
Philippine Airlines	Pal Express	2008	2010	2
Qantas	Impulse Airlines	2000	2004	4
Royal Air Maroc	Atlas Blue	2004	2009	5
SAS	Snowflake	2002	2004	2
United Airlines	Ted	2004	2009	5
	Shuttle by United	1994	2001	7
US Airways	MetroJet	1998	2001	3

include the following (CAPA 2008; Edwards 2008; Gross and Schröder 2007 and Sismanidou, Palacios and Tafur 2008):

1. Common and modern aircraft types.
2. Ancillary revenues/no frills/low frills.
3. Strategic flight scheduling:
   - Flying to cheaper, less congested secondary airports;
   - (mostly) limitation to short- and medium-haul (continental flights);
   - simplified route structures, emphasizing point-to-point services instead of transfers at hubs → no connecting flights (internal or interline);
   - fast turnaround times → allowing maximum utilization of aircraft.

4. Personnel policy, e.g. workload at maximum of legal feasibility/'all-rounder' staff.
5. A simple fare scheme and low fares, including very low promotional fares
6. Unreserved seating (or reservation for a fee), single class configuration and high seating density.
7. Emphasis on direct sales of tickets, predominant usage of Internet-based booking.
8. Strong simple branding that equates low cost with high value, clear consumer message focused on price.
9. Avoidance of competition on same routes (in particular within the LCC sector).

Common and Modern Aircraft Types

Closer examination revealed that 68 per cent of the analysed airlines use only one type of aircraft, while 26 per cent use two types, and 8 per cent use more than two different aircraft types (see Figure 1.2). Out of the 74 airlines using only one type, 33 LCCs chose Boeing B737 variants, 32 LCCs deploy aircraft of the Airbus A320 family; and nine airlines utilize a uniform fleet of another type (e.g. A330/A340, B767/B757, BAe Jetstream 41, Bombardier CRJ, Dash 8, ATR42/72, Embraer 170/190, Fokker 100, MD80 series, or Saab 340).

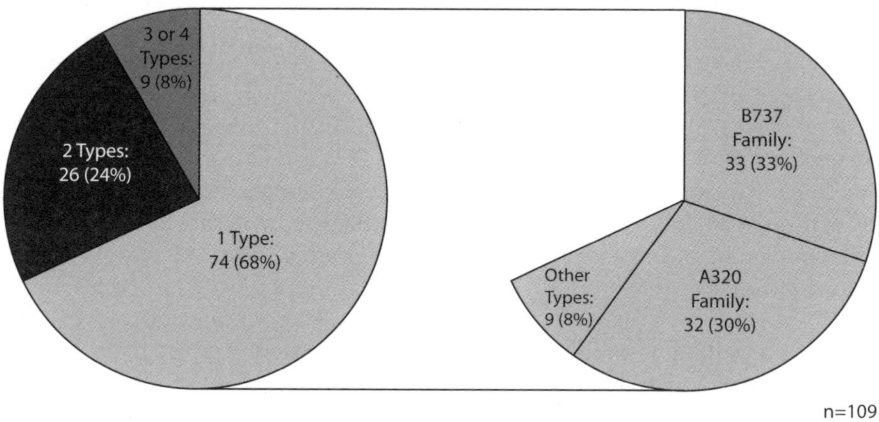

Figure 1.2 Number of aircraft types utilized by low cost carriers

Single Class Configuration and a Simple Fare Scheme

A similar situation can be observed within another typical element of the low cost business model.

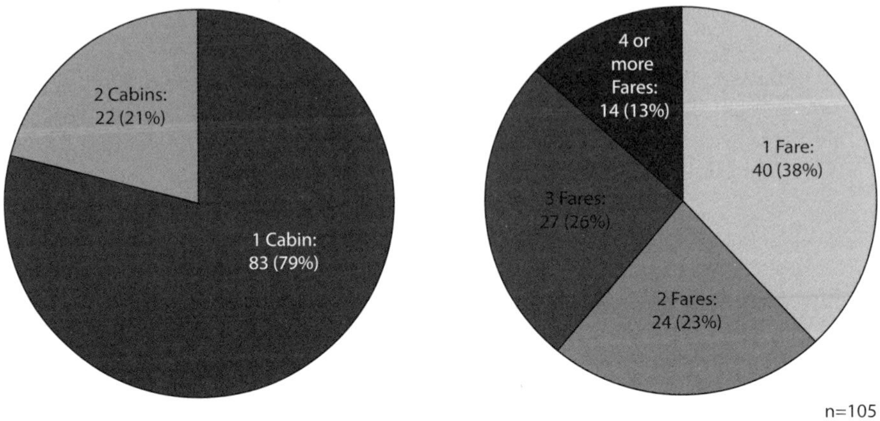

Figure 1.3 Percentage of the utilization of different cabins

The majority of LCCs offer only one cabin class (economy class), while approximately 20 per cent of LCCs include a premium cabin (premium economy or business class). A total of 38 per cent of the all-economy carriers offer one fare, while 13 per cent of the airlines indicate four or more fares (Figure 1.3).

Ancillary Revenue

Within the context of an emerging business model of LCCs, the additional revenue or ancillary revenue of airlines has gained significant importance over the last few years. The consulting company IdeaWorks estimates the 2010 worldwide ancillary revenue within the airline industry to be about €18 billion (Sorensen 2010). This additional revenue is a key pillar of the low cost business model (e.g. in 2009 Ryanair had €663 million in additional revenue; Southwest Airlines approximately €323 million; and Air Asia approximately €230 million) and can be classified as shown in Figure 1.4.

More interesting than the absolute amount of additional revenue is the relative proportion of ancillary revenues, and even more striking is its relation to the return on sales and the profit margin[5] (see Figure 1.5). The leaders in both relative amount of additional revenue and return on sales are the airlines Air Asia and Allegiant.

5 The profit margin is derived from the ratio of operating profit to total profit. Compared to net income, operating profit is more appropriate for an international comparison, as impacts based on different tax systems or other effects are minimal.

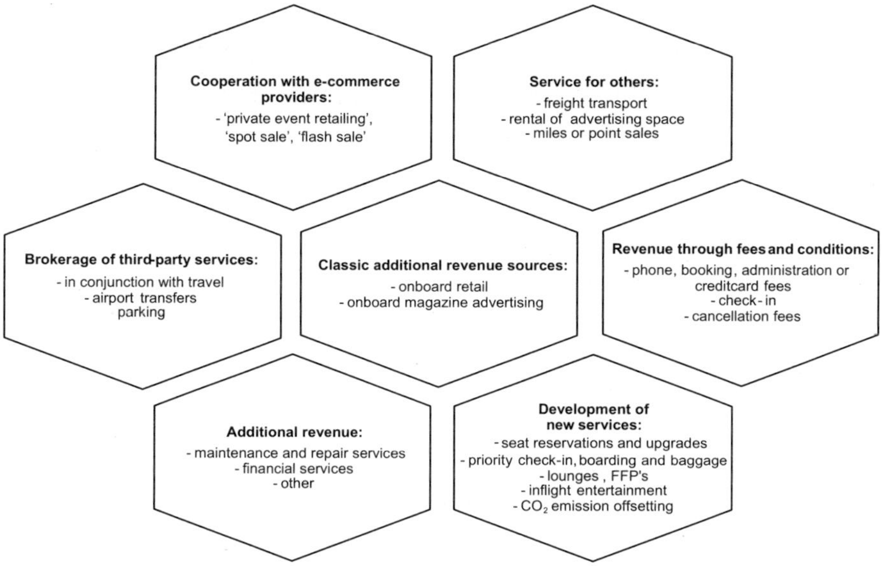

Figure 1.4 Categories of low cost airline ancillary income
Source: Schröder and Freyer 2010.

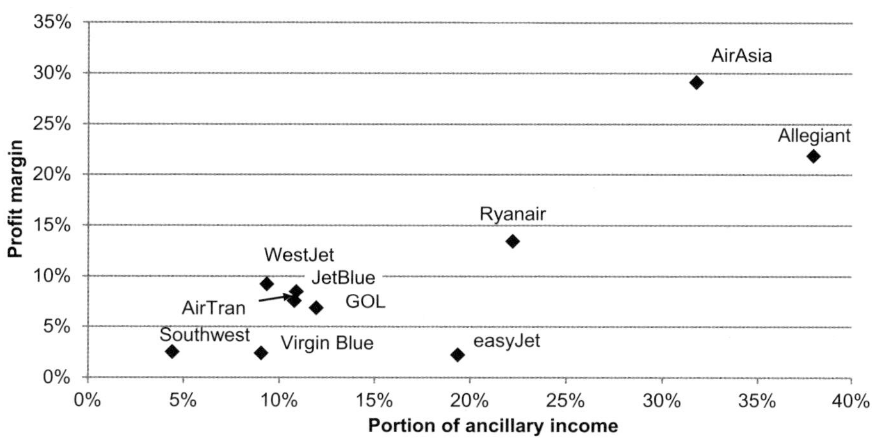

Figure 1.5 Profit margins and ancillary income of low cost carriers (2009–2010)[6]
Source: Schröder and Freyer 2010.

6 For the analysis of LCCs in terms of their additional income, the selection criterion was the publication of annual reports for at least the past five years and the airline size (revenue, passengers). These figures were based on the publication *Airline Industry Guide* 2009/10 (Airline Business 2010). The following airlines have a financial end of year that

Air Asia and Allegiant Air have focused their business model heavily on the achievement of additional revenue and have been economically very successful with this strategy. Southwest Airlines' relatively small proportion of ancillary revenue can be partially attributed to the fact that the first two pieces of luggage are included in the airfare. Although Ryanair generates the highest ancillary revenue in absolute terms among the analysed airlines, the carrier 'only' came in third position within this comparison.

Conclusion

Based on the generated data a number of interesting developments within the global low cost market were highlighted. While more than 200 LCCs existed, or were planned, almost half of those airlines are either no longer active or never started operations. Within this analysis, it was noticeable that the European aviation market was markedly different in several ways.

There was a comparatively high number of withdrawals from the market, and yet currently Europe has the second largest number of LCCs after Asia. A further examination of the number of passengers also shows that LCCs in Europe carry the greatest number of low cost passengers.

Concerning the classic elements of low cost business models, it was demonstrated that there is no unique low cost business model, and that there is some variation among those elements described by some as typical for LCCs.

References

Aberle, G. (2003). *Transportwirtschaft* (4th ed.). Munich, Germany: Oldenbourg.

Airline Business (2009). Airline Industry Guide 2009/10, 10. Sutton, UK: Flightglobal.

Airline Business (2011). Airline Industry Guide 2010/11, May, vol. 11. Sutton, UK: Flightglobal.

Air Transport Intelligence (2008). Low cost carrier profiles. Sutton, UK: Flightglobal.

Bailey, E.E., Graham, D.R. and Kaplan, D.P. (1985). *Deregulation the Airlines*. Cambridge, MA: MIT Press.

Barrett, S.D. (1997). The implications of the Ireland–UK airline deregulation for an EU internal market. *Journal of Air Transport Management*, 3(2), 67–73.

Bjelicic, B. (2004). Osteuropa – Wachstumsmarkt für low cost airlines. *Internationales Verkehrswesen*, 7/8, 309–13.

---

differs from the calendar end of year: easyJet: September 2009; Ryanair, March 2010; Virgin Blue: June 2010.

Harbison, P. and McDermott, P. (eds) (2009). *Global LCC Outlook Report*. Sydney, Australia: CAPA Centre for Asia Pacific Aviation, 266.

Edwards, J. (2008). What is a low cost airline? Defining carrier business models, in *Proceedings of the Air Transport Research Society (ATRS)*, World Conference 6–10 July 2008, Athens, Greece.

ELFAA (2011). *Current Airline Membership*. Brussels: European Low Fares Airline Association. Available at: http://www.elfaa.com/members.htm [accessed: 12 March 2012].

Flight International (2010). World Airlines. Sutton: FlightGlobal.

Francis, G., Humphreys, I., Ison, S. and Aicken, M. (2006). Where next for low cost airlines? A spatial and temporal comparative study. *Journal of Transport Geography*, 14, 83–94.

Gross, S. and Schröder, A. (2007). Basic business model of European low cost airlines, in Gross, S. and Schröder, A. (eds), *Handbook of Low Cost Airlines – Strategies, Business Processes and Market Environment*. Berlin: Erich Schmidt Verlag, 31–50.

Knorr, A. (2007). Southwest Airlines: The low cost pioneer at 35, in Gross, S. and Schröder, A. (eds), *Handbook of Low Cost Airlines – Strategies, Business Processes and Market Environment*. Berlin: Erich Schmidt Verlag, 77–109.

Pompl, W. (2007). *Luftverkehr – Eine ökonomische und politische Einführung* (5th ed.). Berlin: Springer-Verlag.

Ryanair (2011). About us. Available at: http://www.ryanair.com/en/about [accessed: 12 March 2012].

Schröder, A. and Freyer, W. (2010). Bedeutung von zusatzeinnahmen bei Airlines – eine geschäftsmodellbasierte Betrachtung, Handout von DGT-Tagung, DGT-Kongresss 2010 Chur, 25–7 November 2010, Chur (unpublished Conference Handout).

Sismanidou, A., Palacios, M. and Tafur, J. (2008). Challenges faced by European low cost carriers in their market growth strategies and emerging revenue growth formulas: Evolution after the revolution, in *Proceedings World Conference of the Air Transport Research Society (ATRS)*, 6–10 July 2008, Athens, Greece.

Sorensen, J. (2010). Airline ancillary revenue soars to €18.4 billion ($22.6 billion) worldwide in 2010. Press release, IdeaWorks. Available at: http://www.ideaworkscompany.com/press/2010/PressRelease54ARReportDistribution.pdf [accessed: 14 October 2011].

PART II
Europe

Chapter 2
Low Cost Carriers in Western and Central Europe

Roland Conrady

Introduction

After North America, Europe is the second continent to be exploited by low cost carriers (LCCs). These airlines have previously unseen dynamics in the Western and Central European air traffic markets. Following radical structural change, LCCs have since become a firmly established part of the air traffic scene in Western and Central Europe. There are no generally recognized boundaries between Europe's regions. Unless explicitly stated, this article regards the following 20 countries as belonging to Western and Central Europe: Finland, Sweden, Norway, Denmark, Iceland, United Kingdom, Ireland, the Netherlands, Belgium, Luxembourg, Germany, Austria, Switzerland, France, Spain, Portugal, Italy, Greece, Malta and Cyprus. Andorra, the Vatican City, Monaco, Liechtenstein and San Marino will not be analysed separately as they have no airports of any relevance to LCCs. Chapter 3, in this volume, takes a look at the other Central and Eastern European countries.

Economic Importance of the Travel and Tourism Industry

The travel and tourism industry plays a highly important part in the national economies of Western and Central Europe (see Table 2.1). This is where the travel and tourism industry – which includes all companies offering travel as their core business purpose – produces services worth US$622.8 billion and employs 7.8 million people. Including the indirect economic effects, the tourism and travel industry accounts for the production of goods and services worth $US1,699 billion, it employs 20.1 million people. This makes the travel and tourism industry responsible for around ten per cent of economic output and one of the largest sectors of industry in Western and Central Europe.

The air traffic segment accounts for a substantial share of economic output. According to an analysis carried out by the International Air Transport Association (IATA), the aviation industry is responsible for a production volume amounting to

Table 2.1 Economic importance of the travel and tourism industry in the 20 countries of Western and Central Europe in 2010

Travel and tourism industry, 2010 estimates		**Per cent of total**
GDP (US$ millions)	622,800	3.55%
Employment (1,000 jobs)	7,804	4.26%
Travel and tourism economy, 2010 estimates (including indirect economic effects)		**Per cent of total**
GDP (US$ millions)	1,698,753	9.67%
Employment (1,000 jobs)	20,131	11.06%

Source: World Economic Forum (2011); internal analyses.[1]

US$331 billion and provides 4.2 million jobs in Europe (including direct, indirect and induced effects) (IATA 2008).[2]

Economic Importance of Air Traffic

Accounting for a share of 39 per cent, the aeroplane is the second most important means of transport for Europeans in 2010. This is the highest level ever reached by the aeroplane as a means of transportation (by comparison, 44 per cent of Europeans went on holiday by car or motorcycle in 2010) (European Commission 2011a). 2009 saw only 35 per cent of Europeans use the aeroplane as a means of transport to take them on their main holiday. However, significant differences can be observed among the various countries (see Table 2.2).

Air travel in the countries of Western and Central European generated the flight-passengers figures shown in Table 2.3. The number of passengers from the 27 member states of the European Union (EU) amounts to 751.059 million (Eurostat 2011).

The volume of air traffic within Europe accounts for 624.9 billion in revenue passenger kilometrres (rpk). By 2029, air traffic volume will rise to 1,409.1 billion in rpk. This is equivalent to an annual growth rate of 4.1 per cent. In 2009, 13.8 per cent of the world's air traffic originated from Europe. This share will fall to 11.2 per cent by 2029 (The Boeing Company 2010).[3]

---

1 The World Economic Forum uses data from Tourism Satellite Accounting, a worldwide annual analysis conducted by World Travel and Tourism Council (WTTC), based on the United Nations World Tourism Organisation (UNWTO) concept.

2 It must be borne in mind that IATA's definition is very generous and covers a vast area: in addition to the 27 member states of the EU, IATA includes further non-EU countries such as Turkey, Russia and a few CIS states.

3 It must, however, be remembered that Boeing's definition of Europe differs from the one described at the beginning. Boeing also considers the Eastern European countries

Table 2.2 The aeroplane as mode of transport for main holidays in 2009

Malta	83%	Italy	38%
Ireland	78%	Belgium	38%
Cyprus	74%	Netherlands	36%
Norway	63%	Spain	36%
United Kingdom	63%	Germany	35%
Denmark	50%	Iceland	30%
Sweden	49%	Portugal	29%
Luxembourg	48%	France	25%
Finland	46%	Switzerland	n/a
Austria	39%	Greece	n/a

Source: European Commission 2010.

Table 2.3 Air passengers in Western and Central Europe in 2009 (in millions)

1.	United Kingdom	198.532	11.	Sweden	25.219
2.	Germany	158.150	12.	Portugal	24.104
3.	Spain	148.318	13.	Austria	21.817
4.	France	117.562	14.	Belgium	21.314
5.	Italy	101.824	15.	Denmark	20.860
6.	Netherlands	46.479	16.	Finland	13.829
7.	Switzerland	35.298	17.	Cyprus	6.730
8.	Greece	32.882	18.	Malta	2.919
9.	Norway	27.674	19.	Iceland	1.837
10.	Ireland	26.269	20.	Luxembourg	1.535

Source: Eurostat 2011.

The Low Cost Market

Historic Development of the Low Cost Carrier Segment

Although the LCC market segment – as we know it today – is still a relatively young phenomenon, a few precursors in the form of Loftleidir Icelandic, Laker Airways and People Express, were flying over 30 years ago (see Figure 2.1).

as well as Turkey as being part of Europe. Airbus has calculated comparable values (Airbus 2010).

Figure 2.1 Precursors of Europe's current low cost carriers

1960 – Loftleidir Icelandic: Forgotten Pioneer
As early as start of the 1960s, the Icelandic carrier took advantage of a relaxed bilateral between Iceland and the USA to operate low fare transatlantic services between Reykjavik and New York, using single-class Douglas DC-6Bs at first but introducing DC-8-63CF jets later. Even though it had to charge full IATA tourist rates for connections to Europe, the overall fare from European cities to the USA via Iceland was still 16 per cent cheaper than those charged by IATA airlines. The service proved highly successful, raising complaints from other carriers, especially SAS, which exerted enough pressure to receive authority in 1963 to offer a low-fare DC-7 service at equivalent rates.

1971 – Laker Airways Visionary ideas
Freddie Laker´s announcement at London´s Savoy Hotel on 30 June 1971 of his proposed 'Skytrain' no frills transatlantic service between London Gatwick and New York marked the start of a long and acrimonious battle against the authorities and established airlines on both sides of the Atlantic. Laker was eventually able to inaugurate its 'Skytrain' service to New York in September 1977 with Douglas DC-10s, later adding Los Angeles and Miami. Even more ambitious plans to add more than 600 Skytrain routes to European destinations failed to gain approval, but paved the way for today´s proliferation of low fare short-haul airlines. The economic recession, low capitalization and increased competition finally put Laker Airways out of business on 5 February 1982. Sir Freddie sued several airlines for conspiracy to put him out of business … and won!

1981 – People Express Transatlantic inspiration
The founder of the new York-based airline took his inspiration from Sir Freddie Laker, beginning low-cost air travel in 1981 from Newark to several local destinations, before starting flights from London Gatwick in 1983. Despite collecting derisive nicknames for its Spartan, no-frills service, the route proved an instant success, leading to the later addition of Montreal and Brussels. Passengers were charged a fee of 3 US-Dollar for each checked bag, and modest amounts for food and beverages, including the 'snak-pak', comprising cheese, crackers and salami. The aggressive acquisition of several local airlines placed an unsustainable financial burden on the airline, however. It ceased to exist on 1 February 1987, after being bought by Texas Air, which assumed all debts and merged the operation into Continental Airlines.

Source: Airline Business Interactive 2011.

The dynamic rate of growth recorded by LCCs in Western and Central Europe has largely been made possible through the liberalization of European air traffic. This is also based on the initial deregulation measures implemented in the USA: 'The initial impetus for change came from the strong public pressure for deregulation of domestic air services in the United States, culminating in the 1978 Deregulation Act' (Doganis 2006: 32).

The intra-Community air traffic market was gradually liberalized through what were known as liberalization packages that came into effect on 1 January 1988, 1 November 1990 and 1 January 1993. Cross-border air traffic within the EU has been extensively deregulated since 1 January 1993. Restrictions on air services offered by providers from other EU countries were completely lifted on 1 April 1997. Today, the market for air transport within the European Union is marked by four basic principles (Sterzenbach, Conrady and Fichert 2009, Doganis 2006 and European Parliament 2000a, 2000b):

- *Entitlement to receive an operating licence*
  Any company that meets specific market entry criteria (including, e.g. proof that it is a European company in the sense of ownership and property rights) is entitled to be granted an operating licence for the provision of air services within the Community. Thereby no distinction is made between regular and occasional services.
- *Entitlement to receive a route licence*
  Any company with an operating licence is entitled to intra-Community traffic rights, including cabotage. Restrictions are only applied if airports or airspace are unable to cope with traffic volume or if environmental problems are given. Under certain conditions, member states can also impose public service obligations as well as market access restrictions for limited periods.
- *Free pricing*
  Every carrier is allowed to set its own prices. Tariffs are automatically approved if they are filed on time with the air traffic authorities. Tariffs can only be rejected at a later date if they are considered to be too high or too low.
- *Application of general competition law*
  The provisions on the protection of fair competition (ban on restrictive practices in the 2002 treaty that established the EC, control of state aid, merger control) must also be applied in general to air transport.

The legal framework has enabled Europe's LCCs to offer new routes at significantly lower prices in all European countries. Ryanair and easyJet in particular have made extensive use of the options available within European liberalization and expanded on a massive scale into European countries outside their home territories of Ireland and the UK respectively.

This way, the European low cost airlines (LCAs) have evolved into pan-European airlines by building up home bases in several European countries.[4] In 2011, Ryanair had 27 home bases in eight Western and Central European countries: Ireland, the UK, Sweden, Germany, Belgium, France, Italy and Spain.

---

4 See also the interactive route map of the European LCCs under http://www.low-cost-airline-guide.com/en/airline-route-map.htm [accessed 23 March 2013].

> One of the biggest drivers of the low-cost carriers has been the internet, enabling airlines to sell directly to the consumer without the distribution costs network carriers faced through the GDSs. Since the first online tickets were sold in 1995, web-sales have grown exponentially across the industry as a whole. By 2007 more than a third of tickets were sold online. While online sales continue to drive European and North American low-cost carrier sales, ironically a number of budget carriers – seeking to tap business traffic and markets where web-sales penetration remains low – have now turned to GDSs to sell their content as part of a dual strategy. (Airline Business Interactive 2011)

All told, the intensity of competition in Europe has increased – also because of LCC expansion (Dobruszkes 2009). In addition, in the wake of EU enlargement in 2004 (from 15 member states to 25), LCCs were seen to be quick in penetrating the new EU markets. A study from 2005 revealed that traffic between Germany and the new EU countries has rapidly developed into a LCC domain (Conrady 2004, OAG 2006, see also Chapter 3)

Market Size

Europe's largest countries also provide the highest volume of low cost flights. The size of the low cost market in Western and Central Europe is shown in Table 2.4.

Table 2.4 Low cost carrier starts, seats and routes in Western and Central Europe, data for a January week in 2011, both directions of a route are contemplated

Country	Starts	Seats	Routes	Country	Starts	Seats	Routes
UK	7,501	1,007,684	824	Sweden	542	89,291	96
Italy	4,689	780,491	666	Portugal	486	83,831	88
Spain	4,244	757,089	659	Belgium	454	80,042	75
Germany	4,096	632,117	507	Denmark	386	60,510	56
France	1,748	286,215	299	Finland	184	22,124	38
Norway	1,348	223,207	159	Greece	66	10,713	17
Ireland	1,342	230,948	143	Malta	63	11,352	24
Switzerland	817	125,539	100	Cyprus	35	6,165	10
Netherlands	703	110,393	110	Iceland	21	3,129	7
Austria	564	81,645	90	Luxembourg	n/a	n/a	n/a

Source: ADV and DLR 2011.

The ten largest European LCCs by passenger numbers are listed in Table 2.5.

Table 2.5 Low cost carriers in Europe in 2011

LCC	Country	Launch	Starts*	Seats*	Routes*	Passengers (millions)	Current fleet	Orders
Ryanair	Ireland	1991	7,695	1,454,355	1,593	72.7	268	34
easyJet	UK	1995	6,630	1,050,628	779	49.7	174	53
Air Berlin	Germany	2002	2,825	439,212	312	33.6	105	82
Norwegian	Norway	2002	1,970	322,505	281	13.0	57	54
Vueling	Spain	2004	1,291	232,380	123	11.0	38	–
Thomson Airways	UK	2004	n/a	n/a	n/a	11.0	55	8
Wizz Air	Hungary	2003	1,180	212,400	332	9.6	30	105
Pegasus Airlines	Turkey	1990	n/a	n/a	n/a	8.6	29	23
Germanwings	Germany	2002	1,032	154,748	137	7.7	30	8
Anadolu Jet	Turkey	2008	n/a	n/a	n/a	5.3	22	–

Note: * Details are in each case a week in January; outward and return flights are taken into consideration.
Source: *Airline Business* May 2011, ADV and DLR 2011. Please note that Airline Business's low cost carrier financial data do not include low cost carrier Flybe.

Table 2.6 Low cost carrier market shares in European air traffic

	2005	2006	2007	2008	2009	2010
Seat capacity market share	21%	24%	28%	31%	33%	35%
Passenger market share	24%	26%	31%	34%	36%	38%

Note: In the study by York Aviation based on these figures, representative flight-schedule weeks from the winter and summer flight plans were analysed using OAG data. The low-cost carriers under review were: easyJet, Flybe, Jet2.com, Norwegian Air Shuttle, Ryanair, Sverige Flyg, Transavia.com, Vueling, Wizz Air (ELFAA Members) and Sterling, bmibaby, Air Berlin, Monarch Airlines, Volare S.p.a., SkyEurope, Centralwings, Germanwings, Flyglobespan. The traffic territory examined covered Northern, Southern, Western and Eastern Europe.

Source: Brass 2011.

The market shares of LCCs in European air traffic have constantly grown over the past few years, see Table 2.6.

Occasionally analyses with higher market shares can be found. Brass determined a passenger oriented LCC market share of 43 per cent in the point-to-point flight segment. This value coincides with estimates from Doganis who, basing them on OAG data, calculated a share of low cost seats on domestic and intra-European routes of 44 per cent. The low cost shares of Western and Central European countries are shown in Table 2.7.

The 20 largest low cost airports based on flights per week are (sorted by size): London–Gatwick, London–Stansted, Dublin, Barcelona, Madrid, Rome–

Table 2.7 Share of low cost seats on domestic and intra-European routes in each country (September 2010)

Spain	65%	Netherlands	29%
United Kingdom	60%	Sweden	29%
Ireland	48%	Austria	26%
Italy	46%	Denmark	26%
Germany	41%	Finland	26%
Norway	40%	France	25%
Portugal	38%	Switzerland	25%
Belgium	31%	Greece	20%

Source: Doganis 2011.

Table 2.8 Market share of low cost carriers at selected Western and Central European airports (2009)

London–Stansted	82.9%	Manchester	45.1%
London–Luton	72.2%	Berlin–Tegel	41.4%
Alicante	68.5%	Brussels	36.6%
Birmingham	59.6%	Barcelona	32.5%
London–Gatwick	53.3%	Dusseldorf	31.2%
Cologne–Bonn	52.6%	Oslo–Gardermoen	27.9%
Palma de Mallorca	52.0%	Milan–Malpensa	27.6%
Malaga	51.1%	Madrid–Barajas	14.6%
Edinburgh	47.9%	Amsterdam–Schiphol	14.3%
Dublin	46.6%	Munich	13.4%

Source: Eurocontrol, quoted in Statista.

Fiumicino, Oslo, Milan–Malpensa, Edinburgh, Amsterdam, Geneva, London–Luton, Riga, Berlin–Schönefeld, Berlin–Tegel, Manchester, Cologne, Milan–Orio, Düsseldorf, Palma de Mallorca (DLR 2011a).[5] The market shares that LCCs account for in European airports are presented in Table 2.8.

Over many years, LCCs have been able to achieve above average market growth. However, since 2007 growth in Europe has levelled off noticeably, see Figure 2.2.

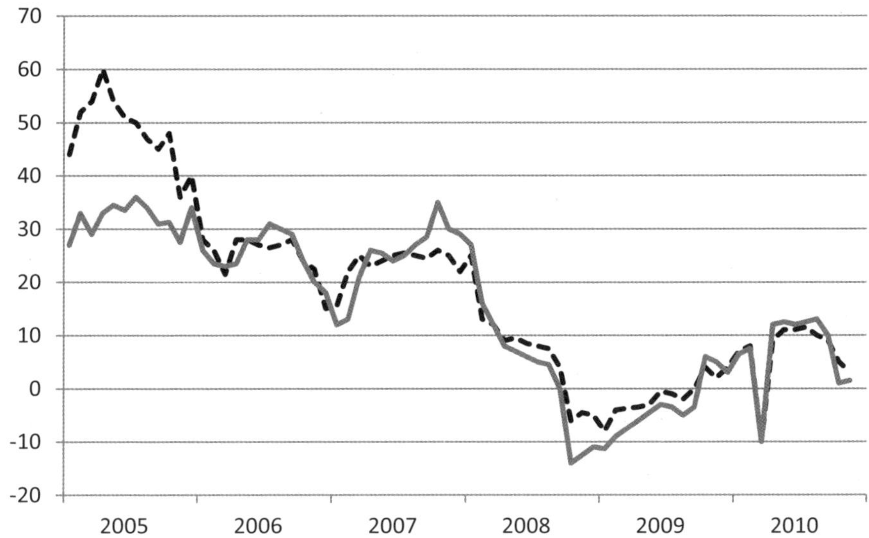

Figure 2.2 Growth rates for the low cost segment (starts and landings, growth rate in % over the same period the year before) in Europe (grey curve) and Germany (black dotted curve)

Source: Eurocontrol, quoted in *Deutsche Flugsicherung* 2011.

Traits and Needs of Low Cost Customers

Only a few studies have been carried out into the socio-demographic and psychosocial traits of low cost passengers in the Western and Central European market. Remarkable are the results of TNS Infratest's Low-Cost Carrier Monitor.[6]

5 Details for a January week in 2011.

6 Since 2005, the Low Cost Carrier Monitor has been performed in Germany, France, Great Britain, Italy and Spain. Every study consists of approximately 11,000 interviews, over 5,000 of which include low cost flight users (Abler and Ehlting 2007, Mason and Alamdari 2007, Schröder and Freyer 2011, Widmann 2007).

> If consumer behaviour and target groups in the European low cost carrier segment are more closely examined, [...] it becomes obvious that low cost flight users exhibit many traits associated with the smart shopper consumer profile. [...] Socio-demographic aspects confirm that low cost flight users do not purchase bargain tickets out of necessity: They have a quite high educational level and an income which is higher than the average air passenger, all of which sets them above the social average. In addition, they often fly more than average – once again confirming the theory that low cost flight products have contributed more to intensifying the flying frequency of existing customers or even, in the worst case scenario for operators, to a cannibalization of high-priced products than to notably opening up the airline market to completely new target groups. More specifically, the passion for brand and quality set the low cost flight user and the smart shopper apart from the simple bargain shopper. As such, the low cost flight user most definitely places value on the airline he uses. Over half of the low cost flight users (51.4 %) in the most important European markets, Great Britain, Germany, France, Italy and Spain, consider the specific airline brand important or even very important when selecting a low cost ticket. In contrast, only every 20th user claims that the airline brand is completely irrelevant for their travel selection. Alongside many established brand airlines, a series of newcomers in the low cost carrier segment are able to enjoy a positive brand image. However, the latter impose some radical convenience and service restrictions on their customers. Disregarding this fact, the airlines are considered safe and reliable and benefit from the image of a cheeky and innovative price breaker, who defied the established airlines and flipped the price structure in favor of the consumers. Last but not least, research and booking behavior confirm parallels to smart shoppers. The large majority of low cost flight users are well-versed with Internet use, enjoy organizing their trips themselves and invest a lot of time and energy in finding the most attractive offer. Many low cost flight users research various relevant offers and, in the course of a value-oriented reflection process, decide which of the offers presents the largest benefit in terms of price, performance, service and brand. These sometimes difficult decisions nevertheless offer an additional attraction: they require above-average market awareness and the ability to interpret and evaluate complex relationships. (Abler and Ehlting 2007: 42)

Development Trends in Europe's Low Cost Market

Bottoming-out of Growth

In some European markets (see Table 2.7), LCCs have already attained a – very high – market share in excess of 40 per cent. This market share is not expected to increase much further.[7] In these markets at least, this means that the LCC segment is likely to stabilize at the growth rates forecast for air traffic in Western and Central Europe of around 3 per cent (Airbus 2010 and Airline Business 2011).

It is anticipated that costs will increase for a number of LCC types. In many cases, airports have offered 'market launch prices' with a view to attracting LCCs. Airport charges in the EU are being scrutinized on an increasing scale under the aspects of state-aid law.

Inside the EU, inclusion of air traffic in the Emissions Trading Scheme (ETS) is expected to be concluded by 2012. Given the particularly price-sensitive nature of their clientele, LCCs face a higher threat from buying emissions certificates than network carriers do. Some European countries also impose special charges on air traffic. In 2010, for instance, Germany decided to introduce an 'environmental departure tax' amounting to €8, €25 or €45 on all of the country's outbound flights. Rising kerosene prices in particular – doubtless to be expected in the future – will hit LCCs harder than it will network carriers. Beyond the avenues of hedging fuel prices, there is nothing the airlines can do to dampen their kerosene costs. LCCs and network carriers must pay the same kerosene prices and face equally high kerosene costs if their fleets use comparable amounts of fuel. A hike in the price of kerosene reduces the percentage price gap between network and LCCs. This results in LCCs losing their appeal. In addition, higher prices lead to more potential air travellers turning their backs on LCCs than on network carriers, either not travelling at all or choosing alternative means of transport. LCC customers are particularly sensitive to prices and respond to increases accordingly. Consequently, LCAs are expected to lose customers to network carriers (Sterzenbach, Conrady and Fichert 2009, Belobaba 2009 in relation to airline cost structures).

Intermodal competition is also expected to grow. Expansion of Europe's high-speed rail networks is also improving the quality of railway services as an alternative product (European Commission 2011b).

Relatively speaking, the above trends will weaken the LCCs' competitive edge. It is hard to imagine that LCCs will be able to achieve above-average sales volumes in the future. The German Aerospace Centre (DLR – Deutsches Zentrum für Luft- und Raumfahrt) in spring 2011 reported declining flights in the low cost segment in its Low Cost Carrier Monitor (DLR 2011b).

Differentiation and Hybridization of Low Cost Business Models

The archetypical LCC business model configures resources and practices that enable airlines to cut costs in comparison to traditional full service airlines (FSAs). This LCC business model as described in textbooks on air transportation includes the following elements (Doganis 2010, Holloway 2008, Sterzenbach, Conrady and Fichert 2009 and Chapter 1 in this volume):

---

7 The low price segment has not been seen to account for more than 50 per cent in other markets either (e.g., food, clothing, etc.).

- point-to-point network;
- a single type of aircraft (usually Airbus 320 or Boeing 737 family);
- a single-class cabin with high seat density;
- predominant use of so-called secondary airports;
- direct sales of tickets, especially over airline's own website;
- no frills such as complimentary in-flight services or frequent traveller programmes; and
- only one one-way fare per flight available at each point in time.

However, in practice this archetypical LCC business model is blurring (e.g. Bell and Lindenau 2009, *DLR Magazine* 2008, Emboaba 2009, Tarry 2010). In Europe, traditional FSAs have joined the battle for cost-conscious short-haul passengers, thereby forcing many European LCCs to change or enhance their business strategy. Additionally LCCs are facing increased competition from other LCCs. Moreover they strive to develop the market of business travellers. Low LCCs are increasingly adopting a hybrid business model modifying key characteristics of their basic business model (for example, Vueling, see Figure 2.3).

An empirical study conducted by the Competence Centre Aviation Management at Worms University examined 25 carriers categorized by DLR in 2009 as LCCs,

Figure 2.3 Vueling is a Type II LCC: Hybrid carriers with dominating low cost elements

Source: Michael Lück.

Table 2.9 Low cost carrier index of European low cost carriers

Type	Airline	Value
I	*Pure low cost carrier*	
	Ryanair	1
	Corendon	0.93
	Jet4you	0.8
	Wizz Air	0.77
II	*Hybrid carriers with dominating low cost elements*	
	bmibaby	0.73
	easyJet	0.73
	Blu Express	0.7
	Iceland Express	0.7
	Jet2	0.67
	Blue Air	0.63
	Aer Lingus	0.53
	Vueling	0.53
III	*Hybrid carriers with dominating traditional airline elements*	
	Air Italy	0.47
	Transavia	0.47
	Wind Jet	0.47
	Germanwings	0.43
	Niki	0.43
	Norwegian	0.43
	Flybe	0.33
	Meridiana fly	0.3
IV	*Traditional full-service airlines*	
	Air Berlin	0.23
	Baboo	0.17
	Blue1	0.17
	airBaltic	0.13
	Brussels	0.1

Note: Value 1 = pure low cost carrier, value 0 = traditional full-service airline.
Source: Conrady, Fichert and Klophaus 2010.

plus four traditional FSAs (British Airways (BA), Iberia (IB), Alitalia (AZ) and Lufthansa (LH)). Examination addressed:

- the extent to which carriers today blend low cost characteristics with the business characteristics of traditional FSAs; and

- which characteristics remain distinct between LCCs and traditional FSAs, and which tend to be common to all carriers.

For the first time a yardstick for positioning the low cost airline business model between the two poles of pure LCC airline and traditional FSA was provided: the LCC index. The study shows that only a few European LCCs are true low cost carriers. Low cost airlines in Europe are often referred to as hybrid carriers dominating low cost elements, hybrid carriers with dominating traditional airline elements, or traditional FSAs, see Table 2.9 on the previous page.

By the same score, traditional FSAs have for some time taken on a number of elements of the low cost business model. They are also making large efforts to cut costs in all company areas. An approximation of the two business models can be observed from this direction too (also refer to Dennis 2007). Consequently, the gap has narrowed in the cost per seat mile available between network carriers and LCCs. This has never been narrower than it is today (also refer to Airline Economic Analysis by Oliver Wyman, quoted in Dunn 2011).

Tapping the Business Traveller Group

For some time now, LCCs have been trying to make their way into the attractive business travel market. The efforts mentioned above towards hybridizing this segment aim at leaving no stone unturned in opening up this target group to the LCC industry. For instance, if easyJet flew into Munich airport, or set up a presence in the global distribution systems, it would be faced with significantly higher costs that would only be justified if easyJet could take a cut of the lucrative revenue from business travel. It is endeavours like these that also explain the structure of frequent flyer programmes that LCCs occasionally offer.

This is why a none-too-insignificant percentage of passengers nowadays use low cost flights for business trips. A survey in 2007 among passengers using Frankfurt–Hahn airport – mainly served by Ryanair – revealed that some 24 per cent of passengers were business travellers (Preissner 2008). The British Civil Aviation Authority (CAA) reported figures of 16.7 per cent for London–Gatwick, 18.8 per cent for London–Luton and 19 per cent for London–Stansted (Tarry 2010).

Tapping the business traveller target group works very well in times of economic recession. It is in time like these when business travellers tend to use more LCCs in an effort to cut costs. Travel guidelines are forming part of business travel management policy in major companies which are initiating cost-cutting measures of this nature. US analyses provide empirical evidence of this trend (Neal and Kassens-Noor 2010). This makes it hardly surprising that times of recession are seeing hardly any change in the operating profit being made by LCCs, contrasted with a dramatic dip in the operating profit of network carriers. This makes the LCC business model look extremely sound (Dunn 2011).

Increasing Route Lengths

It can be observed that since 1994 the average LCC route lengths in Europe have increased (SH&E 2007). This trend can be clearly recognized when considering the current flight schedules for easyJet, which recently opened its longest route (London–Heathrow–Amman); and Ryanair, which is increasingly flying longer Mediterranean routes (e.g. Agadir; Morocco).

Thus LCCs penetrate further into the network carrier market. However, the competitive advantage LCCs have against network carriers is reduced with increasing flight distance. Therefore it is not to be expected that the increase of flight distances presents an attractive expansion opportunity for LCCs.

Figure 2.4 Air Berlin joined the oneworld alliance in 2012
Source: Michael Lück.

Market Structure Changes

In terms of consolidation the air transport market structure will change for two reasons: the competitive ratio of low cost and network carriers will be differentiated; and the market cleansing will continue.

So far, low cost and network carriers have stood in direct competition, which is influenced by a competitive relationship. In addition selective approaches to cooperation will be found in the future. Thus network carriers will – again[8] – acquire holdings in LCCs, just like Lufthansa German Airlines has done with Germanwings.[9] Codesharing and interlining partnerships will also be increasingly realized (see Iberia – Vueling [Figure 2.3] and Air Berlin – oneworld [Figure 2.4]). Network carriers will also increasingly use LCCs as feeders for long-haul flights for cost reasons.

Network carriers as well as LCCs are expected to shake out the aviation market of Western and Central Europe. Competition among LCCs will intensify. With the compression of LCC route networks more and more routes will be served by several LCCs. This leads to yield pressure (Paolo, Stefano and Redondi 2010), which weaker LCCs will not be able to endure. The current number of LCCs will fall. At the same time weak network carriers will have to withdraw from the market or undertake downsizing. Stronger LCCs will make use of resulting additional gaps in route networks.

Conclusion

The LCC business model is considered to be future-proof, because 'any business with an absolutely low-cost base, supplying a product that is demanded by the market, will always be a winner' (Tarry 2010: 30).

A further growth of the LCC market share in Europe is expected for the next few years even though the growth of low-cost business models in Western and Central Europe has diminished. According to Brass (2011) the market share of LCCs in 2020 will be between 45 and 53 per cent in scheduled intra-European passenger traffic.

References

Abler, G. and Ehlting, M. (2007). Smart shopping in the European low cost carrier segment, in Conrady, R. and Buck, M. (eds), *Trends and Issues in Global Tourism 2007*. Berlin, Heidelberg: Springer-Verlag.

ADV and DLR (2011). Low Cost Monitor 1/2011: Der aktuelle Low Cost Carrier markt in Deutschland. Frühjahr 2011. Cologne: Deutsches Zentrum für Luft- und Raumfahrt (DLR).

8 By the turn of the millennium LCC network already had subsidiaries: BA had GO and KLM Buzz.

9 See also Air France's stake in Transavia or the takeover of AirTran by Southwest Airlines in the US Outside Europe it can be observed that LCCs are slowly converging to network carriers: 'JetBlue, WestJet and GOL have stepped up their co-operation with network carriers, Virgin Blue is embarking on alliances with network players Air New Zealand, Etihad – and regulators permitting – Delta' (Dunn 2011: 34).

Airbus (2010). *Airbus Global Market Forecast 2010–2029*. Toulouse, France: Airbus SAS/EADS.

Airline Business (2011). Airline Industry Guide 2010/11, May. Sutton, UK: Flightglobal, 25.

Airline Business Interactive (2011). *The Essential Guide to Low Cost Carriers*. Available at: www.flightglobal.com/iablowcost [accessed: 5 April 2012].

Bell, M. and Lindenau, T. (2009). Sleepless nights. *Airline Business*, 15(3), 70–72.

Belobaba, P.P. (2009). Airline operating costs and measures of productivity, in Belobaba, P., Odoni, A. and Barnhart, C. (eds), *The Global Airline Industry*. Chichester, UK: John Wiley, 113–51.

Boeing Company (2010). Boeing current market outlook 2010–2029. Available at: http://www.boeing.com/commercial/cmo/ [accessed: 12 May 2012].

Brass, J. (2011). *Market Share of Low Fares Airlines in Europe*. Final Report, York Aviation for ELFAA. Brussels: European Low Fares Airline Association, February.

Conrady, R. (2004). *Perspektiven der EU-Erweiterung für die deutsche Tourismusindustrie*. Unpublished study by order of the Federal Association of the German Tourism Industry (BTW) and the International Tourism Trade Show in Berlin (ITB Berlin), Berlin.

Conrady, R., Fichert, F. and Klophaus, R. (2010). *European LCCs Going Hybrid: An Empirical Survey*. Research paper prepared for the 14th Air Transport Research Society ATRS Annual Conference, 6–9 July 2010, Porto, Portugal.

Dennis, N. (2007). End of free lunch? The responses of traditional European airlines to the low-cost carrier threat. *Journal of Air Transport Management*, 13(5), September, 311–21.

Deutsche Flugsicherung (2011). Luftverkehr in Deutschland: Mobilitätsbericht 2010. Langen, Germany.

DLR Magazine (2008). Airline business models. Köln-Porz, Germany: Deutsches Zentrum für Luft- und Raumfahrt e. V. (DLR).

DLR (2011a). *Low Cost Monitor 1-2011*, November, Deutsches Zentrum für Luft- und Raumfahrt e. V. (DLR). Available at: http://www.dlr.de [accessed: 11 November 2011].

DLR (2011b). *Keine weiteren Zuwächse im Low Cost-Segment des Luftverkehrs – mögliche Folge der neuen Luftverkehrsteuer*. Deutsches Zentrum für Luft- und Raumfahrt e. V (DLR). Available at: http://www.dlr.de [accessed: 11 November 2011].

Dobruszkes, F. (2009). Does liberalisation of air transport imply increasing competition? Lessons from the European case. *Transport Policy*, 16(1), 29–39.

Doganis, R. (2006). *The Airline Business* (2nd ed.). London, New York: Routledge.

Doganis, R. (2010). *Flying Off Course: Airline Economics and Marketing* (4th ed.). New York: Routledge.

Doganis, R. (2011). *Responding to Crisis: The Changing Structure of the Airline Industry*. Presentation at the Airneth Annual Conference, The Hague, 14 April 2011.

Dunn, G. (2011). Growth expectations. *Airline Business*, 27(7), 28–36.
Emboaba, M. (2009). Watch out for hybrids. *Airline Business*, 25(3), 57.
European Commission (2010). Survey of the attitudes of Europeans towards tourism. *Flash Eurobarometer 291 – The Gallup Organization*, Analytical Report, Wave 2, March 2010. Brussels: European Commission.
European Commission (2011a). Survey of the attitudes of Europeans towards tourism. *Flash Eurobarometer 328 – The Gallup Organization*, Summary Wave 3, May 2011. Brussels: European Commission.
European Commission (2011b). *High Speed Rail Network*. Available at: http://tentea.ec.europa.eu/en/ten-t_projects/ten-t_projects_by_transport_mode/rail_includes_ertms.htm [accessed: 5 April 2012].
European Parliament (2000a). *European Parliament Fact Sheet – Air Transport: Market Access*. Available at: http://www.europarl.europa.eu/factsheets/4_5_5_en.htm [accessed: 5 April 2012].
European Parliament (2000b). *European Parliament Fact Sheet – Air Transport: Competition and Fares*. Available at: http://www.europarl.europa.eu/factsheets/4_5_6_en.htm [accessed: 5 April 2012].
Eurostat (2011). Eurostat Statistics Database, European Commission. Available at: http://www.epp.eurostat.ec.europa.eu [accessed: 12 May 2012].
Holloway, S. (2008). *Straight and Level: Practical Airline Economics* (3rd ed.). Aldershot, UK: Ashgate.
IATA (2008). *The Economic and Social Benefits of Air Transport 2008*. Geneva: International Air Transport Association.
Mason, K. and Alamdari, F. (2007). EU network carriers, low cost carriers and consumer behaviour: A Delphi study of future trends. *Journal of Air Transport Management*, 13, 299–310.
Neal, Z.P. and Kassens-Noor, E. (2010). The business passenger niche: Comparing legacy carriers and Southwest during a national recession. *Journal of Air Transport Management*, 17(4), 231–2.
OAG (2006). *European Low-Cost Carriers White Paper*. Luton, UK: OAG.
Paolo, M., Stefano, P. and Redondi, R. (2010). *Low Cost fares' Response to New Entry*. Research paper prepared for the 14th Air Transport Research Society ATRS Annual Conference, 6–9 July 2010, Porto, Portugal.
Preissner, U. (2008). Unpublished presentation on Frankfurt-Hahn airport, May.
Schröder, A. (2010). *Das Phänomen der Low Cost Carrier und deren Beeinflussung raum-zeitlicher Systeme im Tourismus*. Erlangen, Germany: Geograph. Ges
Schröder, A. and Freyer, W. (2011). Bedeutung von Zusatzeinnahmen bei Airlines – eine geschäftsmodellbasierte Betrachtung, in *Innovationen in Tourismus und Freizeit*. Berlin, 245–58.
SH&E (2007). *Low Cost Carriers – Interiors*. Presentation by McKenzie, B. at the 9th Annual Aircraft Interior Design & Cost Management Conference, Nice, France, 10 October 2007.
Statista (2010). Anteile der Low-Cost-Carrier an den gesamten Starts auf ausgewählten europäischen Flughäfen in 2009. Available at: http://de.statista.

com/statistik/daten/studie/154957/umfrage/anteile-der-starts-von-low-cost-carriern-auf-flughaefen-in-europa/ [accessed: 11 May 2011].

Sterzenbach, R., Conrady, R. and Fichert, F. (2009). *Luftverkehr: Betriebswirtschaftliches Lehr- und Handbuch* (4th ed.). Munich, Germany: Oldenbourg Verlag.

Tarry, C. (2010). Low-cost commodity. *Airline Business*, 6(2), 28–30.

Widmann, T. (2007). The contribution of low cost carriers to incoming tourism as exemplified by Frankfurt-Hahn Airport and the Rhineland Palatinate Destination of the Moselle Region, in Gross, S. and Schröder, A. (eds), *Handbook of Low Cost Airlines – Strategies, Business Processes and Market Environment*. Berlin: Erich Schmidt Verlag, 171–83.

World Economic Forum (2011). *The Travel & Tourism Competitiveness Report 2011*. Geneva and New York: World Economic Forum. Available at: http://www.weforum.org/issues/travel-and-tourism-competitiveness/

Chapter 3
Low Cost Carriers in Eastern Europe

Borislav Bjelicic

Sub-markets and Aviation Policy Environment

Defining 'Eastern Europe' and the countries belonging to this region is a challenge. Depending on whether you use historical, political, linguistic or cultural criteria, the resulting cluster of countries is bound to differ. For the purposes of this article, and for simplification, the East European market is grouped into four large sub-markets. The first sub-market comprises the East European countries that joined the European Union (EU) in two rounds: Poland, Hungary, the Czech Republic, Slovakia and Slovenia and the three Baltic states of Estonia, Latvia and Lithuania (together known as the 'EU-8') were admitted in 2004, followed by Bulgaria and Romania in 2007. A second group of countries make up the second sub-market: the Balkan countries of Albania, Bosnia-Herzegovina, Macedonia, Croatia, Serbia, Montenegro and Kosovo. Belarus, Moldova and the Ukraine form the third group, with the Russian Federation being the fourth sub-market.

The EU's common aviation market can be seen as one of the great achievements of the European unification process. The EU aviation market was deregulated and liberalized in several stages during the period from 1987 to 1997 (see Chapter 2). As a result, airlines domiciled within the EU nowadays no longer face any regulatory restrictions regarding the way they structure their offers for journeys within the EU: they are free to determine routes, capacity and pricing. The only restriction is being imposed by antitrust authorities whose job it is to ensure competition is working and is not impeded by the actions of commercial enterprises. The East European countries which joined the EU in 2004 and 2007 assumed all rights and obligations that were binding on EU member states at the time of joining (the so-called 'acquis communautaire').

Creating a common aviation market was a key prerequisite for the market entry of numerous new carriers, particularly in the low cost segment. Essentially, their innovation was to increase seating capacity on board their aircraft – by reducing seat pitch between seats – as well as radical cost management across their entire operational process chain. Consumers benefited in the form of constantly falling ticket prices. Of course, this had been precisely the intention of those politicians who had fought for deregulation and liberalization within the EU. At the outset of this development, low cost carriers (LCCs) were easily differentiated against the so-called 'legacy carriers'. Nowadays, however, the differences have significantly diminished, reflecting the profound changes legacy carriers have undergone

(Bjelicic 2007). Thanks to various measures taken to contain costs, legacy carriers have caught up with LCCs,[1] and are nowadays often competitive in their pricing. Legacy carriers market remaining seat capacity in a targeted manner, or generate additional demand through special offers for selected routes. The most obvious difference between legacy carriers operating in Europe and LCCs is the fact that the latter still predominantly operate pure point-to-point connections (even though there are now some exceptions). Compared to network connections through hubs, point-to-point routes have the advantage of being independent of any feeder flights, and are thus not exposed to the related risk of delays. In contrast, legacy carriers are traditionally network carriers, particularly those serving intercontinental routes.

Motivated by the intention of achieving better air transport offers for consumers beyond the confines of the EU, the European Commission initiated negotiations with several Balkan states (see the second sub-group detailed at the beginning of this chapter) and with Iceland and Norway back in 2005, with the aim of establishing a European Common Aviation Area (ECAA) and the objective of transferring the EU regulatory framework in aviation to this area. A multilateral treaty establishing such an ECAA was signed with the West Balkan countries in 2006 (EU Commission 2008).[2] This has already led to significantly higher air traffic, accompanied by an increase in the number of routes and carriers, including LCCs.

At the end of 2007, the EU commenced official negotiations with the Ukraine, with the aim of integrating the country into the ECAA. These negotiations were still ongoing in 2012; once concluded, it is fair to expect additional low cost offers to emanate. Initial progress in liberalizing the market has already become visible: for instance, LCC Wizz Air (see Figure 3.1) already serves several routes to the Ukraine. Aviation policy negotiations with Moldova commenced in 2011, and in June 2012, the EU and Moldova signed a comprehensive air services agreement, which will remove all restrictions on prices and number of flights between the EU and Moldova (European Commission 2012). The talks with Ukraine and Moldova are part of the EU's 'European Neighbourhood Policy' working programme.[3] While Belarus is also a neighbour of the EU, no tangible progress has been achieved due to ongoing political tensions.

The aviation policy relationship with the Russian Federation is particularly important to the EU, since Russia represents the biggest national air transport market across Eastern Europe. However, this relationship has been rather tricky during recent years, with fees charged to EU carriers for overfly rights en route

1 Results for the US in particular have shown a strong decline in the cost differential between low cost carriers and legacy carriers over recent years (Hazel et al. 2011).

2 Bulgaria and Romania were not yet EU members at the time; however, they participated in these negotiations and also signed the treaty. By virtue of their EU accession in 2007, EU aviation law was henceforth directly applicable to these two countries.

3 Reports on the progress made are available at http://ec.europa.eu/world/enp/documents_en.htm [accessed 6 March 2012].

Figure 3.1 Wizz Air serves several routes to the Ukraine
Source: Michael Lück.

to Asia one of the key areas of disagreement. Moreover, Russia has refused to insert an 'EU designation clause' into its existing bilateral aviation agreements with EU member states for quite some time.[4] After close to five years of gridlock, the first long-awaited EU–Russia Aviation Summit finally took place in October 2011, with the EU Commission hoping to find a solution to these disputed issues. Using Twitter, EU Transport Commissioner Siim Kallas reported directly from the conference: 'I am pleased that Russia has today confirmed its full acceptance of EU designation in aviation' (12 October 2011), and 'Minister Levitin confirmed that Siberian overflight royalties will be phased out with Russian accession to the WTO' (13 October 2011). But additional problems have surfaced lately. Being one

4 'Bi-lateral air service agreements between an individual Member State and a non-EU country have to include an "EU designation clause" recognizing that the terms apply equally to all EU airlines, and not just the airlines of that Member State. This is an essential part of the Single European Aviation Market which was created in the early 1990s, guaranteeing that airlines are entitled to operate under the same conditions anywhere in the EU. The requirement to have an "EU designation clause" was confirmed in the "Open Skies rulings" of the Court of Justice in 2002. The Court stated that provisions limiting the benefits of air service agreements to nationals of the Member State concerned are in breach of EU rules on freedom of establishment (now laid down in Article 49 of the Treaty on the Functioning of the EU). Most agreements with non-EU countries have since been adapted to the Court ruling' (European Commission 2010).

of the opponents to the introduction of Emissions Trading Scheme (ETS) Aviation, the EU emissions trading system for the aviation industry, Russia has joined other countries threatening to take countermeasures should their airlines be made subject to the ETS, which started in January 2012. As a first countermeasure, Russia refused to grant new overflight rights to Finnair in June 2012 (Reuters 2012).

But even if the dispute between the EU and Russia over the ETS were to be resolved soon, it will probably take quite some time for the EU and Russia to agree upon a liberalization of their air traffic relationships, akin to the EU/US open skies policy or inclusion in the European Common Aviation Area.

Development of the LCC Market in Eastern Europe

Development of Supply

Since the two rounds of EU accession in the years 2004 and 2007, numerous new air traffic routes have been launched from/to Eastern Europe. Many of those were the result of an expansion of LCCs: overall, 16 LCCs were active on routes from/to Eastern Europe in the summer timetable of 2012. The majority of carriers are domiciled in Western Europe, with only three internationally active LCCs having their corporate head office in Eastern Europe: Wizz Air (established 2003, Hungary), Blue Air (2004, Romania), and airBaltic (1995, Latvia, see Figure 3.2)

Figure 3.2 airBaltic is Latvia's low cost carrier
Source: Michael Lück.

(data from *Flight International* 2010). Two additional carriers based in Russia have only been active in the domestic Russian market so far, but both ceased operations in 2011 (see below). While Wizz Air and Blue Air were founded as LCCs, airBaltic is an example for a strategy shift, from a legacy carrier towards a LCC model. In 2010, airBaltic won a special award from the Jury of the Budgies World Low Cost Airline Awards, recognizing its achievements as a hybrid airline (airBaltic 2010). The European low cost market saw not only many new market entrants over recent years but also some carriers leaving the market. One of the larger insolvencies concerned an East European LCC, SkyEurope. Established in 2001, the airline commenced its flight operations from Bratislava airport and quickly grew, propelled not least by a successful Initial Public Offering (IPO) in 2005. However, in 2009 the airline had to file for insolvency and suspend its flight operations.

In the summer timetable of 2012, Air Berlin has serviced the highest number of East European destinations (27 – including 11 in Russia) among all LCCs active in this geographical market. Norwegian and Wizz Air serve 25 destinations each. At the start of 2004, Ryanair did not fly to any destination in Eastern Europe, but in the summer timetable of 2012, the airline had covered as many as 21 destinations, including nine in Poland; airBaltic serves 16 destinations, followed by Germanwings with 15 destinations. These include only those routes where the airline uses its own aircraft. On top of this, Germanwings offers further East European routes, operated using equipment of its Lufthansa parent (and some others served by holiday carrier Condor). This reflects Lufthansa's efforts to more closely coordinate its group entities. However, as far as Germanwings is concerned, this approach dilutes its original concept of an independent LCC within the Lufthansa Group.

A decision concerning the future of the Germanwings brand was taken by Lufthansa in December 2012. All of Lufthansa's European routes not operating out of the airline's Frankfurt or Munich hubs will be transferred to Germanwings until July 2013. This includes crews and aircraft. All Germanwings aircraft will get a new livery. The rebranded airline will offer three types of fares, ranging from a budget price for a basic service package to a top end fare for a service package which will resemble the Lufthansa top end services on their European flights out of Frankfurt/Munich (Lufthansa 2013).

In the summer timetable of 2012 easyJet was flying to ten destinations.[5] While easyJet flies to seven East European capital cities and three tourist destinations (Split, Dubrovnik and Krakow), Ryanair only serves four capitals, with its remaining destinations largely belonging to the 'visiting friends and family' category. All other carriers have been flying to eight or less destinations.

The next consolidation is already impending, with bmibaby having ceased operations on 9 September 2012 in connection with the sale of BMI, bmibaby's

5 In early 2004, Prague was easyJet's only destination in Eastern Europe.

parent company, to IAG. The original plan was to find a buyer for bmibaby – alas, without success.

Table 3.1 shows the number of destination airports (ranked by passenger numbers) served by the various LCCs during the summer timetable of 2012. The columns indicate the number of destinations served by each LCC, while the number of LCCs serving a particular airport can be seen in the lines. This shows that Prague and Dubrovnik are the destinations attracting the biggest numbers of LCCs.

The liberalization of the aviation market in the EU and the European Common Aviation Area has brought about a significant increase in supply, whereas Russia is still largely terra incognita for LCCs. Very few of the carriers named above fly to destinations in Russia, and there have so far been only two LCCs active in the domestic Russian market: Skyexpress (established 2006) and Avianova (established 2009). In 2011, both of these airlines, which had been founded with combined Russian and Western capital, had to cease operations due to weak financial results. In terms of passenger numbers, Avianova (1.30 million passengers) and Skyexpress (1.14 million) were ranked 12th and 15th in 2010 among airlines domiciled in Russia.[6] One of Avianova's investors was Indigo Partners, who specialize in LCCs and also hold a stake in Wizz Air.[7] So, compared to other traffic regions, the number of LCCs actually domiciled in Eastern Europe is relatively low at just three (see Table 3.2).

LCC traffic from and to Eastern Europe has increased the pressure on legacy carriers in the region. LCC traffic at Budapest airport, for example, accounted for a 2010 market share of 28.6 per cent in terms of aggregate scheduled flight passengers (i.e. excluding charter flights) (Budapest Airport 2011). Hungary's Malev, which had been privatized in 2007, also suffered from increasing LCC competition and was renationalized in 2010, but finally ceased all flight activities on 3 February 2012 after it had run into increasing financial difficulties. Rumours about a sale of CSA (Czech Republic) and JAT (Serbia) have been around for some time. CSA abandoned its attempt (dubbed 'Click4Sky') to market residual seating capacity in a more targeted manner, as a response to LCC offers, at the end of 2009. In Poland, LOT's attempt to challenge its LCC competitors with Centralwings, a subsidiary founded in 2004, failed: Centralwings was allowed to file for insolvency in 2009. Meanwhile, the market share of LCCs in the entire Polish scheduled air traffic volume (i.e. excluding charter flights) has reached 51.6 per cent. Their market share is still lowest in Warsaw (23.8 per cent), whereas it significantly exceeds the average at all other airports, even exceeding the 90 per cent level in Katowice and Lodz (ULC 2011).

6 TCH Transport Clearing House publishes passenger numbers for Russian airlines; current news on Russian airlines can be found, for example, online. Available at: http://www.ruaviation.com [accessed 6 March 2013].

7 Other investments in LCCs include Spirit (USA), Tiger Airways (Singapore), and Volaris (Mexico).

Table 3.1 Destinations served by low cost carriers flying from/to Eastern Europe (summer timetable 2012)

Airport	No. of pax. 2010 (000's)	FR	GO	AB	DY	VY	W6	EI	4U	BE	HV	LS	WW	BT	IV	0B	AE	Total
Moscow–DME (RUS)	22,253			•		•								•	•			4
Moscow–SVO (RUS)	19,123													•				1
Prague (CZ)	11,556		•		•	•	•	•	•		•	•	•		•		•	11
Moscow–VKO (RUS)	9,460								•									1
Warsaw (PL)	8,666				•		•	•						•			•	5
St Petersburg (RUS)	8,391			•	•	•								•	•			5
Budapest (HU)	8,179		•		•		•	•	•		•	•		•				8
Kiev–Boryspil (UA)	6,700				•									•	•			3
Bucharest OTP (RO)	4,917			•		•	•	•	•						•	•		7
Riga (LV)	4,663	•			•		•							•				4
Sofia (BG)	3,296		•	•			•											3
Krakov (PL)	2,839	•	•	•	•			•	•			•						7
Belgrade (SRB)	2,698			•	•		•		•									4
Ekaterinburg (RUS)	2,697			•														1
Katowice (PL)	2,366	•					•											2
Gdansk (PL)	2,210	•		•	•		•											4
Novosibirsk (RUS)	2,131			•														1
Krasnodar (RUS)	2,093			•														1
Zagreb (HR)	2,071		•		•	•			•									4
Bourgas (BG)	1,894				•		•	•			•							4
Bratislava (SK)	1,665	•			•													2

Table 3.1 Continued

Airport	No. of pax. 2010 (000's)	FR	GO	AB	DY	VY	W6	EI	4U	BE	HV	LS	WW	BT	IV	0B	AE	Total
Wroclaw (PL)	1,598	•					•											2
Tirana (AL)	1,536								•									1
Samara (RUS)	1,535			•														1
Ufa (RUS)	1,460			•														1
Rostov (RUS)	1,438			•											•			2
Ljubljana(SLO)	1,388		•				•											2
Tallinn (EST)	1,384	•	•		•									•				4
Poznan (PL)	1,383	•			•		•											3
Vilnius (LT)	1,373	•			•		•	•						•			•	6
Pristina (XK)	1,305		•	•	•				•									4
Minsk (BY)	1,285													•				1
Dubrovnik (HR)	1,270		•		•	•		•	•	•	•	•	•					9
Varna (BG)	1,227				•		•											2
Split (HR)	1,219		•	•	•	•	•		•		•	•						8
Timisoara (RO)	1,139						•											1
Irkutsk (RUS)	1,039			•														1
Cluj–Napoca (RO)	1,028						•									•		2
Kaliningrad (RUS)	1,024			•										•				2
Kazan (RUS)	958			•														1
Chisinau (MD)	937													•				1
Simferopol (UA)	841						•							•				2

Table 3.1 Continued

Airport	No. of pax. 2010 (000's)	FR	GO	AB	DY	VY	W6	EI	4U	BE	HV	LS	WW	BT	IV	0B	AE	Total
Kaunas (LT)	809	●												●				2
Perm (RUS)	747			●														1
Skopje(MK)	716			●			●											2
Odessa (UA)	707													●				1
Kiev Zhuliani (UA)	700						●											1
Chelyabinsk (RUS)	658			●														1
Sarajevo (BIH)	563				●				●									2
Tivat (MNE)	532			●														1
Lviv (UA)	480						●											1
Rzeszow (PL)	451	●																1
Lodz (PL)	413	●					●											2
Brno (CZ)	396	●					●											2
Pula (HR)	313	●		●	●				●			●						5
Zadar (HR)	272	●		●					●									3
Szczecin (PL)	268	●			●													2
Bydgoszczc (PL)	266	●																1
Bacau (RO)	241															●		1
Sibiu (RO)	198				●											●		2
Palanga (LT)	102				●									●				2
Constanta (RO)	78	●		●														2
Tirgu Mures (RO)	74						●											1

Table 3.1 Continued

Airport	No. of pax. 2010 (000's)	FR	GO	AB	DY	VY	W6	EI	4U	BE	HV	LS	WW	BT	IV	0B	AE	Total
Rijeka (HR)	61	●			●				●									3
Plovdiv (BG)	26	●																1
Debrecen (HU)	24			●														1
Osijek (HR)	20	●																1
Arad (RO)	13			●														1
Satu Mare (RO)	11															●		1
Total		21	10	27	25	7	25	8	15	1	5	6	2	16	6	5	3	

IATA code	Airline	IATA code	Airline	IATA code	Airline	Airport mnemonic
FR	Ryanair	EI	Aer Lingus	WW	bmibaby	SVO = Sheremetyevo
GO	easyJet	4U	Germanwings	BT	airBaltic	DME = Domodyedovo
AB	Air Berlin	BE	Flybe	IV	Windjet	VOK = Vnukovo
DY	Norwegian	HV	Transavia	0B	Blue Air	
VY	Vueling	LS	Jet2	AE	Iceland Express	
W6	Wizz Air					

Sources: Airlines' proprietary analyses and airport websites.

Table 3.2 East European LCCs (July 2012)

Carrier	IATA code	Country	Type of service	Founded	Fleet size	Destinations
airBaltic	BT	Latvia	Scheduled	1995	34	54
Blue Air	0B	Romania	Scheduled	2004	10	25
Wizz Air	W6	Hungary	Scheduled	2003	36	79

Sources: *Flight International* (2010); company websites.

In parallel with air traffic developments, numerous East European airports were expanded during recent years. LCCs benefited in particular from the expansion of secondary airports, which some of them prefer. One of many examples in this context is Sibiu airport in Romania, where the expansion project, which started in 2006, was not least driven by Sibiu's award as Cultural Capital of Europe 2007. In Poland, the capacity of airports in Gdansk, Poznan, Wroclaw and Warsaw was expanded in the wake of the European football championships in 2012.

As in Western Europe, only a few East European cities have more than one airport, Moscow, with Domodyedovo, Sheremetyevo and Vnukovo airports, being the most obvious exception. Air traffic in Bucharest is shared among the Henri Coanda (previously called Otopeni) and Aurel Vlaicu (previously Baneasa) airports. Since the beginning of the summer schedule 2012, Aurel Vlaicu airport – which is near the city centre – is exclusively used by business jets. All the remaining low cost traffic there was shifted to Bucharest Henri Coanda (Romanian Insider 2012). Kiev also has two passenger airports: Borispol International airport (accounting for some seven million passengers) and Zhulyany airport. There are plans to expand Zhulyany, and Wizz Air was the first LCC to relocate there in early 2011 (It's Ukraine 2011). Low cost traffic at other major airports is concentrated in certain terminal areas (such as Terminal 1 in Budapest), or there is no separation (as yet) between LCCs and other international scheduled airlines (in Belgrade, for example). Privatization of East European airports is still in its early days; so far, airports in Budapest, Tirana, Varna, Bourgas and Skopje have been privatized (ACI Europe 2010).

Looking at developments over the past few years, it is fair to say that the Internet has contributed decisively to the commercial success of all the low cost airlines. Low cost carriers recognized the potential of Internet sales for marketing their tickets at a very early stage. Consequently, they have been pioneers for using the World Wide Web for booking and paying for flights. Thanks to the proliferation of Internet usage, LCCs were even able to expand this sales channel. However, there are still regional differences in the penetration of the Internet today. The USA has developed furthest in this regard, with 78 per cent of the population (or an estimated 245 million out of the total population of 313 million) using the Internet (Miniwatts 2012, including the following figures as of 31 December 2011). Looking at the 17 EU member states outside Eastern Europe, the average Internet penetration (also based on the overall population) is 75.3 per cent, against an average rate of 58.8 per cent for the ten East European member states. There are vast differences between the various countries: among East European EU members, Estonia currently has the highest (77.5 per cent) and Romania the lowest (39.2 per cent) Internet penetration. In the two most populous countries, Ukraine and Russia, the current share of Internet users in the overall population only amounts to 33.9 per cent and 44.3 per cent respectively. The differences between Western and Eastern Europe are even greater when considering the use of the Internet for online shopping for goods and services. In 2010, the share of Internet users who were also using the web for online shopping was 79 per cent in the

UK; in Romania this was a mere 9 per cent (Seybert and Lööf 2010). The reasons for this discrepancy include the still low prevalence of credit cards as a means of payment in Eastern Europe. Therefore, alternative distribution channels such as telephone sales are likely to play a role in Eastern Europe for some time.

Especially on routes from/to Eastern Europe, LCCs face another group of competitors rarely seen as prominent: international long haul bus lines, established on numerous routes since the 1990s, when a liberalization of air transport – and the market entry of LCCs – were still unthinkable. Besides operators who have specialized in certain country-to-country connections, there are large network operators such as Deutsche Touring, who were offering routes to 700 destinations in 32 countries in their summer 2012 timetable. Since bus lines serve a much tighter network of places, the distance to the next bus stop is often rather short, compared to air travel. Advantages for bus passengers compared to air travellers therefore often become evident when looking at overall travel times. Moreover, terms and costs for carrying luggage are often more attractive when using the bus. Nowadays, some bus operators already operate pricing models similar to those of low cost airlines, by offering attractively priced seats for certain times of travel. Given that long haul buses have a more favourable carbon footprint compared to air transport, an increasing internalization of external costs might turn out to be beneficial for bus travel (Haunerland et al. 2009). The development of long haul bus lines was additionally supported over recent years by the investment focus of East European countries, which supported the expansion of road infrastructure at the expense of developing their rail networks (ITF 2011).

Demand Trends

The key motivations for using passenger air transport are business travel, holidays, and visiting friends and relatives (VFR). Business and holiday travel are together referred to as 'tourist travel'. When looking at the routes offered by LCCs, it is evident that VFR journeys are the dominating reason for using some routes (e.g. to Lodz, Katowice or Pristina), whereas others (such as Varna or Dubrovnik) are clearly tourist destinations. The connections serving East European capital cities (e.g. Prague, Budapest, etc.) are mainly characterized by a mix of the main travel reasons mentioned. Consequently, more business travellers – who are not the typical customers of LCCs – can be found here.

Migration has been the main driver of VFR journeys from and to Eastern Europe, with work and studies being the key triggers of migration movements from Eastern to Western Europe. Initially, migration set in at the beginning of the 1960s, from Yugoslavia (which was a co-founding member of the Non-Aligned Movement and as such not locked in behind the Iron Curtain), especially towards the Federal Republic of Germany. Citizens of all ex-Yugoslavian countries are still widely present in Germany today. A second wave of migration from other East European countries started after the disintegration of the Eastern Bloc, and with the evolution of democratic societies in the former Eastern Bloc countries. At the time, the disintegration of Yugoslavia, but (periodically) also the desolate state of Albanian society and its economy, led to further migration movements. The last major development was triggered by the two EU accession rounds in 2004 and 2007.

Within the framework of the 2004 accession, Cyprus and Malta were the only two new EU member states to immediately benefit from full freedom of movement for their workers. In contrast, transitional provisions were introduced for the eight East European accession countries, the EU-8. Specifically, the 15 existing EU member states were able to decide on the speed with which they wished to open their labour markets and the transitional periods they wanted to impose on the EU-8. Some of the member states, such as Spain and the United Kingdom, opened their labour markets to EU-8 citizens at a very early stage, whereas Germany made extensive use of the scope for transitional provisions. In any case, the duration of any labour market restrictions for EU-8 countries had to end on 30 April 2011, and full freedom of movement for workers from the EU-8 countries has been in effect since 1 May 2011. As a result, the conditions for migration within the EU once again improved, even though it is fair to assume that there will be no more sharp increases. This is also due to the fact that in the wake of the economic crisis which has been ongoing since 2008, the prospects for immigrants have deteriorated in several countries, such as Spain and Ireland. In fact, there are signs of a weakening of East European immigration to some of these countries. There are also indications of new migration movements from those countries particularly affected by the sovereign debt crisis – Portugal, Spain and Greece – towards Germany and northern Europe. The high unemployment rate among young people is a particular reason for the well-qualified among them to consider migrating to other European countries.

Upon the EU accession of Romania and Bulgaria in 2007, the option to introduce transitional provisions restricting freedom of movement for workers was also established. These were used by various countries to a different extent; full freedom of movement will be established by 31 December 2013. All EU member states are still free to decide individually on immigration rules vis-à-vis East European countries outside the EU.

Looking at the current situation, East–West migration has been concentrated – in absolute figures – on the most populous countries, namely, Germany, France and the United Kingdom, and on Spain and Italy in the Mediterranean region (see Table 3.3).[8] Italy is top of the list, with almost 2.2 million immigrants from Eastern Europe, of which 968,000 were Romanian citizens. Italy was also the main European target country for the migration of workers from Albania (482,000), and

---

8 The figures shown in the table for France relate to the year 2008 (INSEE 2008), the data for the UK depict the period from October 2010 to September 2011 (ONS 2012); the remaining figures were drawn from the EU's Eurostat database and relate to the year 2011 (Eurostat 2011).

Table 3.3 Migration from Eastern Europe to main European Union target countries

(000's)	Destination							
Citizens from	**Germany**	**France**	**Spain**	**Italy**	**UK**	**Ireland**	**Netherlands**	**Total**
Poland	448	92	85	109	654	86	52	1,526
Czech Republic	37	7	9	6	34	7	3	103
Slovakia	28	4	8	9	59	11	3	122
Hungary	73	9	9	7	42	5	6	151
Estonia	5	1	2	1		3	1	13
Latvia	15	2	4	2	67	19	2	111
Lithuania	25	2	23	4	123	35	3	215
Slovenia	21	2	1	3			1	28
Romania	135	50	843	968	84	12	8	2,100
Bulgaria	80	14	171	51	43	1	14	374
Croatia	233	9	2	21			1	266
Serbia	193	67	3					263
Montenegro	14							14
Macedonia	70	4	1	90			1	166
BiH	162	13	1	32			2	210
Albania	10	6	2	482	13			513
Ukraine	136	13	87	201	13	2	3	455
Moldova	13	5	17	131		1		167
Belarus	20	3	4	7			1	35
Russia	206	42	54	30	24	2	5	363
Total	1,924	345	1,326	2,154	1,156	184	106	7,195

Sources: Compiled from INSEE (data for France for 2008), ONS (data for the United Kingdom for 2011), and Eurostat (all other countries for 2011).

also attracted a notable share of Ukrainians (201,000) and Moldovans (131,000). Germany comes second, with 1.9 million East European immigrants, mainly originating from the successor countries of former Yugoslavia (672,000), Poland (448,000) and Romania (135,000). In addition, 136,000 Ukrainian and 206,000 Russian citizens have migrated to Germany. The Spanish population has also grown considerably due to migration over recent years. Immigrants to Spain from Eastern Europe predominantly originated from Romania (843,000) and Bulgaria (171,000). The fact that Spain and Italy have turned out to be so popular among Romanians is due to the close linguistic ties; overall, close to 90 per cent of all Romanian migrants to Europe moved to Spain or Italy.

In the United Kingdom, Polish citizens account for the largest group of foreign nationals by far (654,000). However, the number of citizens from the Baltic countries living in the UK is particularly high compared to other countries. In terms of its overall number of inhabitants, France has seen the lowest migration from Eastern Europe; the main countries of origin are Poland (92,000), Serbia (67,000) and Romania (50,000). Similarly, The Netherlands exhibit a very low share of East European immigrants in terms of the overall population. In contrast, Ireland has a relatively high share of immigrants from Eastern Europe among countries in Western Europe (184,000, or 4.1 per cent of its overall population of 4.5 million). Main countries of origin are Poland (86,000) and the Baltic countries (57,000).

The migration statistics reveal why the main low cost routes to Poland start in Germany and the United Kingdom, while most low cost routes to Romania originate in Spain and Italy. In this context, it is worth noting that migration statistics do not always provide sufficient grounds for estimating potential passenger numbers. This is due to the fact that many migrant workers have acquired citizenship in the country they have migrated to over recent years (Vasileva and Sartori 2008), while their ties to family and friends in their former home country have remained intact and are typically passed on to the next generations.

Of all East European destinations served by low cost airlines, only a few are dominated by tourist travel. The Croatian airports on the Adriatic coast – most notably Dubrovnik – are a case in point. Pula is the northernmost Croatian airport near the coast and it is also important for neighbouring Slovenia's coastal holiday resorts. Due to its short coastline, Montenegro only has a relatively small airport (Tivat). Airports on the Black Sea coast (Varna, Bourgas and Constanza) are also dominated by tourism. Across the continent, only the two capital cities of Prague and Budapest, as well as Krakow in Poland, have established themselves as popular holiday destinations in Eastern Europe. In Russia, Saint Petersburg and Moscow are the main tourist destinations. Besides tourism, business travel has developed favourably over recent years, since EU enlargement has created a host of new business opportunities for companies from both Western and Eastern Europe. This is evident when looking at the increase in direct investments in Eastern Europe.

However, a look at the statistics published by the World Tourism Organization (UNWTO 2006, 2011, 2012, including the following figures) reveals that developments in travel to Eastern Europe failed to match many high expectations during recent years (see Table 3.4). The number of international tourist arrivals

Table 3.4 Tourist arrivals in Europe (2000, 2008, 2010)

Inbound tourists (000's)	2000	2008	Change 2000–2008	2010	Change 2008–2010
Estonia	1,220	1,970	61.5	2,120	7.6
Latvia	509	1,684	230.8	1,373	-18.5
Lithuania	1,083	1,611	48.8	1,507	-6.5
Poland	17,400	12,960	-25.5	12,470	-3.8
Hungary	12,212	8,814	-27.8	9,510	7.9
Czech Republic	4,773	6,649	39.3	8,629	29.8
Slovakia	1,053	1,767	67.8	1,327	-24.9
Bulgaria	2,785	5,780	107.5	6,047	4.6
Romania	867	1,466	69.1	1,343	-8.4
Slovenia	1,090	1,940	78.0	1,869	-3.7
Eastern EU members	*42,992*	*44,641*	*3.8*	*46,195*	*3.5*
Belgium	6,457	7,165	11.0	7,186	0.3
Germany	18,992	24,886	31.0	26,875	8.0
Finland	2,714	3,583	32.0	3,670	2.4
France	77,190	79,218	2.6	77,148	-2.6
Greece	13,096	15,939	21.7	15,007	-5.8
Ireland	6,646	8,026	20.8	6,515	-18.8
Italy	41,181	42,734	3.8	43,626	2.1
Luxembourg	852	879	3.2	793	-9.8
Malta	1,216	1,291	6.2	1,336	3.5
Netherlands	10,003	10,104	1.0	10,883	7.7
Austria	17,982	21,935	22.0	22,004	0.3
Portugal	12,097	6,962	-42.4	6,756	-3.0
Spain	47,898	57,192	19.4	52,677	-7.9
Cyprus	2,686	2,404	-10.5	2,173	-9.6
Denmark	3,535	9,016	155.0	8,744	-3.0
UK	25,209	30,142	19.6	28,299	-6.1
Sweden	2,746	4,555	65.9	4,951	8.7
Other EU members	*292,500*	*328,039*	*12.2*	*318,643*	*-2.9*
Croatia	5,831	9,415	61.5	9,111	-3.2
BiH	171	322	88.3	365	13.4
Macedonia	224	255	13.8	262	2.7
Albania	32	1,330	4056.3	2,347	76.5
Other countries	*6,258*	*11,322*	*80.9*	*12,085*	*6.7*

Sources: UNWTO 2006, 2011, 2012.

(comprising holidaymakers and business travellers) in the ten East European EU member states rose by only 3.8 per cent from 2000 to 2008, compared to a 12.2 per cent increase for the other EU countries during the same period. However, since the onset of the financial markets crisis at the end of 2008, the trend has shifted somewhat towards certain East European member states. While the number of international tourist arrivals in East European EU member states increased by 3.5 per cent from 2008 to 2010, tourist arrivals in other EU member states declined by 2.9 per cent during the same period.

However, there are marked differences among East European countries. Since 2000, the number of international tourist arrivals has dropped significantly in Poland and Hungary, while Bulgaria and the Baltic countries were successful in winning tourists. Since 2000, international tourist arrivals have also increased strongly in Albania and in some of the successor countries to former Yugoslavia; the same holds true for the Ukraine, whereas numbers have stagnated in Russia.

In conclusion, it is worth taking a look at visa regulations, which continue to be an important factor determining travel between the EU and other countries in Eastern Europe. Not long ago, visa requirements were waived for travellers from several Balkan countries to the EU except Ireland and the United Kingdom (for Serbia, Macedonia and Montenegro since December 2009, and additionally for Bosnia-Herzegovina and Albania since December 2010), provided that travellers hold biometric passports. Negotiations are being held between the EU and Moldova, as well as with the Ukraine, to further liberalize existing visa regulations.[9] In this context, the relationships between the EU and Russia – as well as Belarus – are the most difficult, given mutual visa requirements, which is a particular burden for travel from and to Russia. Even though tourist arrivals in EU countries from Russia have developed favourably over recent years, tourist destinations from Russia are still primarily concentrated on countries that do not require a visa from Russian citizens, or where visas can be obtained more easily (e.g. upon arrival) – such as Turkey, Egypt, the United Arab Emirates and Thailand. Overall, however, only a relatively small section of the Russian population has been travelling abroad to date, with Muscovites accounting for more than 40 per cent of that group (ETC 2010).

Outlook

After years of strong growth, favoured by EU air transport policy, the growth momentum of LCCs now appears to be slowing down across the board. This also applies to routes from and to Eastern Europe. The economic and political opening of numerous East European countries triggered a strong migration towards the

9 However, relations between the EU and the Ukraine have been burdened recently by tensions regarding the trial of Yulia Tymochenko, the former Ukrainian Prime Minister, who was sentenced to several years in prison in October 2011.

West, which has since weakened. Likewise, tourism is not expected to yield any strong impulses; this is due to the fact that only few destinations in Eastern Europe are typical (mass) tourism locations. Also, competition from Spain, Italy, Turkey and other Mediterranean countries is too strong.

There is still some potential in terms of city tours and cultural tourism, though: quite a number of East European cities offer interesting architecture and a vibrant cultural scene. Examples include the Ukrainian cities of Lviv and Odessa; there are also Zagreb, capital of Croatia, and Ljubljana, capital of Slovenia. Rising quality standards in infrastructure pertinent to tourism are bound to attract a growing number of conferences and similar events to be held in Eastern Europe. The 2012 European Football Championships held in Poland and the Ukraine provided additional momentum for aviation; similarly the 2014 Winter Olympics in Sochi, Russia. The 2018 Football World Cup will also take place in Russia. As for traffic flows in the opposite direction, growing affluence and higher incomes among private households will create stronger demand in the coming years for air travel from Eastern Europe to tourist destinations in Western Europe. The number of potential travellers is particularly high from Russia – however, existing visa regulations present a burden, especially for individual travel.

Overall, competition is set to increase, and not just among LCCs. On East European routes they will also compete with legacy carriers – even more so as differences in service levels and pricing fade, an effect that is already visible in the saturated North American and West European markets. Major risks exist with regard to the development of jet fuel prices. Given the prevailing income differential compared to Western Europe, customers in Eastern Europe are likely to be significantly more price-sensitive than consumers in the West. On the other hand, the democratization process in Eastern Europe was driven by the desire for freedom, accompanied by the wish to travel freely. Air travel epitomizes this wish better than any other form, which is why the attitude towards air travel is set to remain predominantly positive.

References

ACI Europe (2010). *The Ownership of Europe's Airports*. Brussels: Airport Council International Europe. Available at: http://www.aci-europe.org/component/content/article/37-publications/41-publications.html [accessed: 25 July 2010].

airBaltic (2010). airBaltic wins special award from world's top low cost carrier jury. Press release, 30 September. Available at: http://www.airbaltic.com/public/45252.html [accessed: 6 March 2013].

Bjelicic, B. (2007). The business model of low cost airlines – Past, present, future, in Gross, S. and Schröder, A. (eds), *Handbook of Low Cost Airlines: Strategies, Business Processes and Market Environment*. Berlin: Erich Schmidt Verlag, 11–29.

Budapest Airport (2011). *Budapest Airport Traffic Profile 2010*. Available at: http://www.bud.hu/english/b2b/airlines/documents_and_statistics [accessed: 4 November 2011].

ETC (2010). *Market Insights: Russia*, 11 August. Brussels: European Travel Commission, Market Intelligence Group. Available at: http://www.etc-corporate.org/images/library/ETCProfile_Russia_6-2010.pdf [accessed: March 2011].

European Commission (2008). *Communication from the Commission: Common Aviation Area with the Neighbouring Countries by 2010 – Progress Report (COM(2008) 596 final)*, 6 August 2008. Braunschweig, Germany: DLR/European Commission. Available at: http://eur-lex.europa.eu/LexUriServ/LexUriServ.do?uri=COM:2008:0596:FIN:EN:PDF [accessed: 22 November 2011].

European Commission (2010). *Air Transport: Commission Launches Infringement Procedures Against France, Germany, Austria and Finland Over Agreements with Russia on Siberian Overflights (IP/10/1425)*, 27 July 2010. Brussels: European Commission. Available at: http://europa.eu/rapid/pressReleasesAction.do?reference=IP/10/1425&format=HTML&aged=1&language=EN&guiLanguage=en [accessed: 19 November 20011].

Eurostat (2011). *Population by Sex, Age Group and Citizenship (migr_pop1ctz)*. Brussels: European Commission. Available at: http://appsso.eurostat.ec.europa.eu/nui/show.do?dataset=migr_pop1ctz&lang=en [accessed: 4 July 2012].

Flight International (2010). World Airlines 2010. London: Reed Business International/FlightGlobal.

Haunerland, F., Moll, R., von Hirschhausen, C. and Walter, M. (2009). Potenzial des Fernlinienbusverkehrs in Deutschland. *Internationales Verkehrswesen*, 61(4), 115–21.

Hazel, B., Taylor, A. and Watterson, A. (2011). *Airline Economic Analysis*. New York: Oliver Wyman Consulting.

INSEE (2008). *Statistiques détaillées sur la répartition des immigrés par pays de naissance en France en 2008*. Paris, France: National Institute for Statistics and Economic Studies. Available at: www.insee.fr/fr/ffc/figure/immigrespaysnais.xls [accessed: 4 July 2012].

ITF (2011). *Investment and Maintenance in Inland Transport Infrastructure 1995–2008*, 20 July. OECD – International Transport Forum. Available at: http://www.internationaltransportforum.org/statistics/investment/Overview.pdf [accessed: 21 November 2011].

It's Ukraine (2011). Wizz Air Ukraine moves from Boryspol to Zhulyany airport, 19 January. Available at: http://www.itsukraine.com/wizz-air-ukraine-moves-from-boryspil-to-zhulyany-airport.html [accessed: 6 March 2013].

Lufthansa (2013). 'New Germanwings' to become Europe's most modern airline with á la carte travel. Press release. Available at: http://www.germanwings.com/downloads/Press-Release-New-Germanwings.pdf [accessed: 23 March 2012].

Miniwatts (2012). *Internet World Statistics*, as of 31 December 2011. Available at: http://www.internetworldstats.com [accessed: 4 July 2012].

Office for National Statistics (ONS) (2012). *Population by Country of Birth and Nationality Oct 2010 to Sep 2011*, Table 2.3: Estimated population resident in the United Kingdom, by foreign nationality, 60 most common nationalities, 24 May 2012. Newport, UK: Office for National Statistics. Available at: http://www.ons.gov.uk/ons/rel/migration1/migration-statistics-quarterly-report/may-2012/population-by-country-of-birth-and-nationality.xls [accessed: 4 July 2012].

Reuters (2012). Russia turns down extra overflights, 8 June. Available at: http://www.reuters.com/article/2012/06/08/russia-airlines-overflight-idUSL5E8H88W620120608 [accessed: 6 March 2013].

Romanian Insider (2012). Traffic at Bucharest Otopeni airport doubles after low cost airlines move flights, 26 March. Available at: http://www.romania-insider.com/traffic-at-bucharest-otopeni-airport-doubles-after-low-cost-airlines-move-flights-in/53629/ [accessed: 6 March 2013].

Seybert, H. and Lööf, A. (2010). *Internet Usage in 2010 – Households and Individuals*. Eurostat Data in focus 50/2010, 8 August 2010. Available at: http://epp.eurostat.ec.europa.eu/cache/ITY_OFFPUB/KS-QA-10-050/EN/KS-QA-10-050-EN.PDF [accessed: 12 November 2011].

ULC (2011). *Analiza Rynku Transport Lotniczego w Polsce w 2010 roku*, 13 August 2011. Available at: http://www.um.gliwice.pl/bip/pub/zam_pub/1709/z16548.pdf [accessed: 14 November 2011].

UNWTO (2006). *Tourism Market Trends* (2006 ed.). Madrid: World Tourism Organisation.

UNWTO (2011). *World Tourism Barometer* (Interim update, April 2011). Madrid: World Tourism Organisation.

UNWTO (2012). *World Tourism Barometer* (Interim update, May 2011). Madrid: World Tourism Organisation.

Vasileva, K. and Sartori, F. (2008). *Acquisition of Citizenship in the European Union*, Eurostat Statistics in focus, 2008/198, 1 August 2008. Available at: http://www.eds-destatis.de/de/downloads/sif/sf_08_108.pdf [accessed: 21 November 2011].

PART III
The Americas

Chapter 4
Low Cost Carriers in the USA and Canada

Timothy M. Vowles and Michael Lück

Introduction

Over the last 30 years, the United States, considered one of the first if not the first market to witness the emergence of low cost carriers (LCCs), has demonstrated a number of traits that have allowed for the materialization of LCCs and their success as a viable form of commercial air transportation. Among the most significant of these are the geography of the country, an early deregulation of the airline industry, and a large population with enough disposable income to travel. Canada, similar in geography, shares many of these traits.

The Airline Deregulation Act of 1978 took the control of pricing and market selection out of the hands of the United States federal government and placed it under the control of the air carriers. Prior to this move the Civil Aeronautics Board (CAB) was the governmental agency with the power to determine what markets a carrier was able to enter and exit and what fares that carrier could charge in any market. With the passage of the 1978 Act and the ultimate end of the CAB, market forces rather than political forces were the driver behind airline decisions. The deregulation of the industry allowed a wave of new carriers to begin serving customers across the United States, thus beginning the rise (and, more often than not, the fall) of LCCs in the United States. This easing of governmental regulation of air transportation markets was not total. The federal government still controls a number of aspects of the industry having a direct impact on all carriers, including LCCs, among which are the limitation on foreign ownership of domestic airlines and the operation of the air traffic control system.

In contrast to the USA, Canada's deregulation was more gradual, due to the recession in the early 1980s and the government's ownership of Air Canada, the country's largest airline (Howell 2003). In 1985, the government released a White Paper, 'Freedom to Move', which proposed a framework for the gradual deregulation of the aviation sector, and resulted in the 1987 National Transportation Act (Nickerson 2001). During this time, the National Transportation Agency (NTA) replaced the Canadian Transport Commission, and the state owned Air Canada was privatized. Along with the USA deregulation, pricing regulations were abolished, and the now private Air Canada faced stiff competition from Canadian Airlines, as well as a number of smaller carriers and charter airlines such as Wardair and Air Transat.

A large amount of research conducted into the LCC success in the United States looks at a wide range of different topics, including examining the role of deregulation (Morrison and Winston 1995, Goetz and Sutton 1997, Barrett 2004, Francis et al. 2006); competitive response (Morrell 2005, Graham and Vowles 2006); and the various components of a successful LCC model (Windle, Lin and Dresner 1996, Windle and Dresner 1999, Graham 2009). This chapter will look at LCCs in North America a bit differently by examining the geography of the LCC phenomenon in North America, focusing on two different areas: the geography of current LCCs in the United States and Canada, and the geography of cities with LCC service. Finally, the geography of Southwest Airlines is used to underline the importance of this airline as the pioneer in LCC services.

Low Cost Carriers in the USA and Canada

With an expenditure of $US145.1 billion in North America, tourism plays an important role in the economies of the USA and Canada. North America saw a total of 101.7 million visitors in 2011, an increase of 6.6 per cent from the previous year. This equals a share of approximately 10 per cent of the world tourism market (UNTWO 2012). After France, the USA represents the world's second biggest tourist market (62.3 million visitors in 2011), whereas Canada was visited by 15.98 million, and Mexico by 23.4 million. The World Tourism Organization, (UNWTO 2012) projects that the market will increase to 120 million visitors to North America by 2020, and 138 million by 2030.

A report published by the Federal Aviation Administration (FAA) in 2011 finds the direct and indirect impacts of the aviation industry to be enormous. Over 5 per cent of the United States economy, US$1.3 trillion, is attributed to civil aviation generating or supporting over 10 million direct and indirect jobs. The same report finds air transportation responsible for almost US$250 billion being spent on goods and services by air travellers.

There has been a decline in employment in the air transportation sector even as the output continues to increase. This decline, according to the FAA report, is partially attributed to the presence of LCCs. The newer aircraft employed by the LCCs means fewer maintenance concerns and, in turn, fewer maintenance employees per aircraft. Other reasons given for the overall decline in the industry's employment include the outsourcing of jobs within an airline to outside firms, and the substitution of technology for certain tasks, including ticketing and check-in along with the reduction of a number of services provided by the airlines.

The geography of the United States plays a key role in the emergence and success of LCCs in the country. This spatial influence is limited not only to the vast size of the country but also to its political geography; and, on a more local scale, its regional and relative geography have also played an important part in the beginnings and ultimate success of LCCs there.

Covering an area of over 9.5 million square kilometres, the United States is the third largest country in the world by geographic size. Although it has one of the world's most comprehensive road networks, movement across this vast geographic space is time-consuming. The advent of commercial air passenger services alleviated some of the challenges of distance, but it was not until the deregulation of the industry in 1978 and the appearance of LCCs soon after that the challenge of distance started to wane for a large number of Americans who, up until that point, were limited in their access to air travel due to the high cost of flying and the limited options of other forms of transportation throughout most of the country (see Table 4.1 on the next page).

While Southwest Airlines is held as the model of success for LCCs not only in the United States but also around the world, the network trend of LCCs in the United States is geographically different from that of Southwest. The other key LCCs in the United States (annual operating revenue over $US20 million) all operate hub and spoke systems or modified hub and spoke systems. This puts them more in line with traditional airlines as far as network structure is concerned. Two of the largest and oldest LCCs, Air Tran and Frontier, operate hub and spoke systems that put them in direct competition with established legacy carriers. Air Tran focuses operations on Atlanta Hartsfield International Airport, where it is in direct competition with Delta Air Lines. Denver is the hub of Frontier's operations, where it is in direct competition with United Airlines and Southwest Airlines.

The newer LCCs have created modified hub and spoke systems combining the point-to-point system with hub and spoke. The carriers usually focus operations on one or two cities, serving a wide number of destinations from these selected cities. While this sounds an awful lot like a traditional hub and spoke model, the difference is that the cities these carriers focus upon are located on the geographic periphery of the country, limiting the number of feasible connections the carrier can create through its focus airports. Jet Blue focuses its operations on two airports on the east coast, New York's John F. Kennedy Airport and Boston Logan International Airport; over 41 per cent of the carrier's total air passengers either begin or end their trip at these two airports (RITA BTS 2011). The geographic positioning of these two airports limits the number of through spokes the carrier can reasonably offer to north/south passenger flows along the east coast of the United States.

Upstart carrier Virgin America (see Figure 4.1) is much smaller than JetBlue but operates a similar network on the west coast of the United States. The carrier focuses its operations on San Francisco International Airport and Los Angeles International Airport. Since the number of west coast destinations the carrier serves are limited (Seattle, San Diego and Las Vegas), Virgin America concentrates on serving transcontinental markets from San Francisco and Los Angeles International airports, creating a more point-to-point style of network.

Spirit Airlines operates what appears on the surface to be a standard hub and spoke system centring on Fort Lauderdale and Detroit. A closer inspection of the geography of the network reveals something a bit different from other hub and spoke networks operated, one focused on using Fort Lauderdale as an

Table 4.1 Developments in the US low cost market (selection)

Airline	Year of foundation	Withdrawal from the market	Parent/holding, Headquarters
Southwest Airlines	1971	—	Dallas, Texas
ATA Airlines	1973 (as American Trans Air) 2003 as ATA Airlines	2008	Global Aero Logistics Inc., Indianapolis
Spirit	1980 (as Charter One) 1992 (as Spirit)	—	Indigo Partners & Oaktree Capital Management, Miramar, Florida
PEOPLExpress	1981	1987 (integrated into Continental)	People Express Airlines Inc., Newark
Sun Country Airlines	1982	—	Cambria Holdings, Mendota Heights, Minnesota
Tower Air	1983	2000	Private shareholders, Jamaica, New York
Midwest Airlines	1984 (as Midwest Express) 2002 (as Midwest)	2010 (merged with Frontier)	Republic Airways Holdings, Oak Creek, Wisconsin
Independence Air	1989 (as Atlantic Coast Airlines) 2004 (as Independence Air)	2006	FLYi Inc., Loudoun County, Virginia
Valujet	1992	1997 (merged with Air Tran)	Public company, Atlanta
AirTran	1994	2012 (acquired by Southwest Airlines)	Airways Corporation, Orlando
Frontier	1994	—	Republic Airways Holdings, Denver
Allegiant	1998	—	Allegiant Travel Co., Enterprise, Nevada
JetBlue	1999	—	Jet Airways Corporation, Long Island City, New York
USA3000	2001	2012	Apple Vacations, Newtown, Pennsylvania
Song	2003	2006 (reintegrated into Delta)	Delta Air Lines Inc., Atlanta
Ted	2004	2009 (reintegrated into United)	UAL Corporation, Elk Grove, Illinois
Virgin America	2007	—	Black Canyon Capital LLC, Burlingame, California

Sources: Compiled from the airlines' websites.

Figure 4.1 Start-up LCC Virgin America operates predominantly on the west coast
Source: Michael Lück.

international gateway. This international focus and the degree of this concentration on international destinations is what separates Spirit geographically from its competitors. Of the 40 destinations the carrier serves, 25 of them are international destinations; this is by far the largest number of international destinations served by a LCC in the United States. All of the international destinations in the Caribbean, Mexico, Central America, and South America flow through the carrier's gateway hub at Fort Lauderdale international Airport. In the case of Spirit the location of their hub operations on the geographic periphery of the United States changes when the scale is modified to a larger hemispheric scale. When the geographic scale is changed in this manner the Spirit's hub location is centrally located to flow passengers between the United States and the Caribbean, Mexico, Central America, and South America.

Allegiant Air is one of the more spatially unique LCCs operating in the United States on two different fronts, the cities it serves and the spatial division of its network. The carrier's network focuses on providing point-to-point service between a number of smaller airports across the United States, a number of which only have service provided by the carrier, or very limited service from other carriers, generally only to a major hub, and a set number of vacation destinations split between the eastern and western parts of the United States. These destinations include Phoenix, Las Vegas, Los Angeles, and Bellingham, Washington in the west; and Orlando, Fort Lauderdale, Fort Myers, Tampa, and Myrtle Beach in the east. Allegiant splits the country in half in organizing which focus destinations they serve from which origin. Those in the western half of the country only service western focus cities; and those in the eastern half of the only service the eastern focus cities. In the middle part of the country, running roughly through the Mississippi Valley, the cities around Lake Michigan and southern Texas are centrally located enough that the carrier can offer service to at least one focus destination in either half. The carrier's unique geographic approach to service has created a service network completely unique in the United States.

Canada is characterized by similarly vast distances between cities across the country, but differs from the USA in that most of the population lives in the southernmost parts and the sub-Arctic and Arctic areas are only sparsely populated. The population size in the large cities results in main trunk routes between these main centres, with few air services to the smaller cities in the far north. For example, the city pairs Montreal–Toronto, Toronto–Vancouver, Calgary–Vancouver, Calgary–Toronto, and Ottawa–Toronto represent the five busiest routes in the country (Howell 2003).

Following deregulation, Canada has a history of emerging LCCs attempting to compete with Air Canada and subsequently ceasing operations (see Table 4.2). Airlines such as CanJet, Jetsgo, Canada 3000 and WestJet (see Figure 4.2) commenced operations between 1989 and 2001. Some hybrid airlines (charter and scheduled services) also entered the low cost market in selling seats only, for example Royal Airlines, Air Transat and Sunwing. In an attempt to compete

Figure 4.2 WestJet is Canada's second largest airline
Soures: Michael Lück.

Table 4.2 Developments in the Canadian low cost market (selection)

Airline	Year of foundation	Withdrawal from the market	Parent/holding, Headquarters
Air Transat	1987	—	Transat A.T. Inc, Montreal
Canada 3000	1989	2001 (merged with CanJet, and collapsed in November 2001)	No parent, Etobicoke (outside Toronto)
Royal Airlines	1991	2001 (was acquired by Canada 3000)	Royal Aviation, Montreal
WestJet	1996	—	Public company, Calgary
CanJet	1991 relaunch 2002	2001 (merged with Canada 3000, and was relaunched after its collapse. From 2006, charter only)	IMP Group International, Enfield (near Halifax)
Jetsgo	2001	2005	None, Montreal
Tango	2001	2003	Air Canada, Montreal
zip	2002	2005	Air Canada, Calgary
HMY Harmony	2002	2007	HMY Airways Inc, Vancouver
Sunwing Airlines	2005	—	Sunwing Travel Group, Toronto

Sources: Compiled from the airlines' websites.

with these LCCs, Air Canada itself established LCCs, such as Tango and zip. zip was established in 2002 to directly compete against WestJet, with both airlines headquartered in Calgary. Tango, short for 'Tan and Go' (referring to the sun holiday destinations it served), was another subsidiary of Air Canada, established in 2001. Both zip and Tango ceased operations after only two years of service, and were reintegrated into the parent airline in 2003 and 2004, respectively. Royal Airlines started operations in 1991, and was acquired by Canada 3000 in 2001. Canada 3000 in turn got into financial difficulties following the 9/11 events, and merged with CanJet in 2001. The only surviving true LCC in Canada is WestJet and, with approximately 16 million annual passengers, it is now Canada's second largest airline.

The Geography of Destinations in the USA

The maturation of the United States LCC industry has allowed the trend to spread across the entire country, with LCCs having a presence at a wide range of airports across the country. All ten of the top domestic markets in the United States are served by LCCs. Table 4.3 shows the top ten domestic air travel markets in the United States from April 2011 to March 2012. All ten are served by LCCs in addition

Table 4.3 Top domestic city pairs in the USA

Route	Passengers (in millions)
Chicago–New York	3.47
Los Angeles–San Francisco	3.34
Los Angeles–New York	3.14
Chicago–Los Angeles	2.88
Atlanta–New York	2.85
Fort Lauderdale–New York	2.61
Atlanta–Orlando	2.59
Atlanta–Washington DC	2.52
Chicago–San Francisco	2.49
Chicago–Washington DC	2.35

Sources: RITA BTS 2012.

to traditional legacy carriers. Of these cities, only Atlanta and Fort Lauderdale are served by only one airport (government statistics do not aggregate passenger data for Miami, Fort Lauderdale, and West Palm Beach into one metropolitan area. If they did, Fort Lauderdale–New York would take the top spot). The remaining cities are served by a number of airports, which allows LCCs to compete directly in the same markets but not necessarily directly at the same airport.

In terms of domestic enplanements, LCCs have a significant presence (greater than 5 per cent of the enplaned passengers at the airport) in six of the top ten airports. Of the four airports in which no LCC has at least 5 per cent of the market share, three of the metropolitan areas Chicago, Los Angeles and Houston are served by secondary airports where an LCC, or, more precisely, Southwest Airlines, is the dominant carrier. If the total traffic in each of the three metropolitan areas were agglomerated between the primary and secondary airport, Southwest would have nearly 25 per cent of the enplaned domestic traffic in all three of the markets. The fourth airport without a strong LCC presence, Charlotte, is dominated by US Airways and only has LCC service provided by Air Tran to its hub in Atlanta, Orlando, and Baltimore, and by Jet Blue to New York's JFK airport.

An examination of Domestic Scheduled Enplanements ranking for the United States in 2011 shows Southwest as the largest domestic carrier in the country, with Air Tran and Jet Blue ranked seventh and eighth. Both Southwest and Jet Blue had the largest percentage increase in passenger enplanements between the first two quarters of 2010 and the first two quarters of 2011, with 6.3 per cent and 9 per cent respectively. This demonstrates the allure and market strength that LCCs have in the United States. Air Tran saw a slight decrease in enplanements during this same time period, but this can be attributed to Southwest Airlines acquiring the carrier and the beginning steps of integration of the two carriers.

While it is generally assumed that LCCs serve primarily domestic leisure markets, an examination of the top three markets for each of the six focus LCCs shows the major markets for each carrier varies and is a combination of both leisure and business destinations. However, one major pattern does emerge among five of the six carriers: the prevalence of a hub or focus city among the top markets. This is obviously a result of the operating systems employed by the carriers. Southwest is the only carrier where the top three markets all had different origins and destinations. Another emerging geographic relationship focuses on the predominant leisure destinations served by each carrier. Those carriers serving the eastern portion of the United States – AirTran, Spirit, and Jet Blue – have a large share of their leisure traffic focused on Orlando, Florida, and Fort Lauderdale, Florida, while those carriers focusing service on the western portion of the country – Frontier, Virgin America, and Allegiant – have Las Vegas, Nevada, as their primary leisure focus. This does not mean that LCCs from one region do not serve leisure destinations in another, it just means that there is a spatial concentration of leisure destination service.

The types of airports LCCs serve is a particularly interesting element of the LCC model. Part of the original Southwest model showed that a carrier would enter into large metropolitan markets by serving secondary airports. The term 'secondary airport' is a bit vague, as it can mean a variety of different airport types. There are two basic types of secondary airport in North America. The first is an older airport in a large city that has been replaced by a larger newer airport. Examples of this type of airport include Chicago Midway, Houston Hobby, and Dallas Love. The second type of secondary airport is a smaller in terms of overall enplanements, and in close geographic proximity to a larger metropolitan area which is served by a larger airport. Examples of these types of secondary airports include Providence, Rhode Island (Boston), Fort Lauderdale, Florida (Miami), and Oakland, California (San Francisco). The interesting thing about some secondary airports is that they are actually the largest airport in terms of domestic passengers in some metropolitan areas. Examples include Fort Lauderdale over Miami and Baltimore/Washington or both Washington DC Reagan and Washington Dulles. In both of these examples the predominant carrier in terms of market share is Southwest followed by Spirit at Fort Lauderdale and AirTran at Baltimore.

One of the hallmarks of the LCC model is the type of destinations to which carriers begin offering service. At least somewhere within a LCC's network will be destinations designated as vacation or leisure. The general purpose behind this strategy is to attract travellers who are price sensitive and are willing to shop for the lowest fares regardless of schedule, frequent flier programmes, and other amenities offered by major carriers. Initially these types of markets were underserved by the major carriers, as they were more focused on the business traveller and their revenue streams. A geographic analysis of LCC networks shows leisure destinations are important to the general LCC model in the United States but far from the only consideration of when a carrier makes service decisions. Southwest focused on the price-sensitive business traveller, initially in the state of

Texas and then throughout its network as it grew into a nationwide carrier. It was not until the mid-1990s that the carrier expanded its network to serve the leisure-heavy destinations in Florida, although in the early 1980s the carrier entered into what were at the time the predominantly leisure markets of Phoenix and Las Vegas.

A quick examination of carrier route maps shows that of the six LCCs discussed in the previous section, all six have the following leisure destinations in common: Las Vegas, Orlando, and Fort Lauderdale. If Virgin America is removed from the tally because of its relatively new entry, and Allegiant because of its spatially unique operation style, two more leisure destination appear: Tampa and Fort Myers, both in Florida. Southwest serves all of the destinations as well. Leisure destinations are definitely a staple of LCC networks in the United States because of the fare-sensitive passengers found in these markets.

An interesting spatial trend is emerging in the United States concerning LCC service, and that is international service. Aside from Southwest and Allegiant. all of the LCCs discussed provide service in at least two international markets. As was mentioned earlier, Spirit leads the way, connecting Fort Lauderdale to 25 international destinations in Mexico, Central America, South America and the Caribbean. All of the carriers offer service to Cancun and, as would be expected because of geographic proximity, those carriers whose operations are focused in the eastern portion of the United States offer service to more international destinations than the two carriers in the west, Frontier and Virgin America. It is interesting to note that none of the carriers offer service to Canada despite the open skies policy between the two countries, though Spirit markets their Niagara Falls service as Toronto service and their Plattsburgh, New York service as Montreal service.

Two questions emerge: Why international expansion? and, Why to the south? The maturation of the United States air market and the LCC market limit expansion opportunities for carriers within the domestic market. All of the large cities in the United States have LCC service, so there only appear to be two major ways to expand networks: internationally; or Allegiant's way of connecting small markets to leisure domestic destinations. The international liberalization of air transportation plays a significant role in LCC's international expansion. The primary way in which liberalization of international air markets is achieved is through open skies agreements between the United States and other countries bilaterally. Since 1998, the United States has signed open skies agreements with nine Central America/Caribbean countries, coinciding with LCC expansion into the region. As with the Airline Deregulation Act, the basic thrust of open skies agreements is to take the control of markets away from the government(s) and place it in the hands of the operators in terms of scheduling, fares and markets served. Safety and other non-commercial characteristics of the service are still influenced by government.

The expansion southward makes sense geographically for two reasons. The first is the obvious large number of leisure markets in Mexico, Central America, South America and the Caribbean that can be tapped into – much like Southwest expanding service in the 1980s and 1990s – against established major carriers offering higher fares. The second reason is that with the economic middle class growing in some of these destinations, traffic originating in these areas may start to grow the markets even more as travellers want to visit the United States.

The Geography of Destinations in Canada

Canada's geography dictates two main markets for its low cost airlines: the domestic inter-city traffic and the holiday traffic to the sun destinations in the south of the USA and in the Caribbean. All three remaining LCCs operate a mix of charter and scheduled flights, while Air Transat has more characteristics of a charter airline. Sunwing and Air Transat operate flights to various cities in central Europe; however, neither of them is offering any domestic services in Canada. Thus, WestJet is the only true domestic LCC in Canada. A total of 59.8 per cent of its business is attributed to domestic services, and 40.2 per cent to charter and scheduled international services (WestJet 2011).

Despite the fact that Canada's overall population density is low, many regions in the south such as Southern Ontario or greater Vancouver have population densities higher than some European countries. The large size of Canada's north cannot support large human populations, so settlements in the north are small and very fragmented. These geographic realities have a direct effect on Canada's aviation sector, in particular on low cost markets. Due to the low population size in the north, regular air traffic is in most cases not viable, and consequently very expensive. Looking at the route map of Sunwing, the northernmost city served by this airline is Edmonton in the province of Alberta. Air Transat serves the sun destinations from some northern airports, such as Halifax in Nova Scotia and St John's in Newfoundland.

WestJet connects a number of larger towns and cities in the north, including Yellowknife in the Northwest Territories and Whitehorse in Yukon, with the rest of the country. WestJet also operates out of cities in the northern parts of British Columbia, Alberta, and out of Quebec, Newfoundland, Nova Scotia and New Brunswick. WestJet's focus of operations is out of the three main bases across the continent (Vancouver, Calgary, Toronto) with significant traffic out of Edmonton and Winnipeg as well.

Despite the relatively good coverage by WestJet (and competitor Air Canada), WestJet announced the launch of regional flights into smaller towns across the country from 2013 (Putzger 2012, CBC News 2012). While the carrier has not yet announced which airports will be served, they have recently ordered 20 Bombardier Dash 8-Q400 turboprop aircraft for its regional arm, and taken options for another 25 aircraft of the same type (CBC News 2012). The regional airline will be a separate airline with its own operating certificate, but wholly owned by WestJet (WestJet 2011).

Case Study: The Geography of Southwest Airlines

The impact of political geography is evident in the creation and early success of what many consider the original LCC, Southwest Airlines. Southwest Airlines began service in 1971 as an intrastate carrier in the state of Texas serving the Texas triangle of Dallas–Houston–San Antonio, the three largest cities in the state. At this time the Civil Aeronautics Board regulated interstate air travel in the United States, controlling, among other things, the fares airlines could charge and the destinations and routes in which they were allowed to operate. By choosing to just serve the state of Texas, Southwest was able to control fares, frequency of service and market entry and exit. Over the next seven years the carrier developed a route network that covered the entire state of Texas and at the same time built a strong customer following in Texas that, once the Airline Deregulation Act of 1978 was passed, removing route control from the federal government, the airline was able to open up service to destinations outside the state with an already strong customer base.

Southwest's operational system during this time period was a point-to-point service much like other carriers across the country. Southwest pursued this point-to-point, short haul strategy for a couple of reasons: first the carrier was originally restricted to the state of Texas; and second, it allowed the airline to provide high frequency services, thereby maximizing its assets – namely its aircraft and crews. However, once deregulation was passed, the vast majority of United States carriers adopted a hub and spoke system of operation, whereby airports in a select number of cities were used as transfer points for passengers. While the hub and spoke system allowed carriers which operated in this fashion to offer more destinations to their passengers, it also made for less efficient use of resources such as aircraft and employees. Southwest did not alter its operational style after deregulation, keeping the point-to-point service that had worked extremely well for it over the previous seven years.

Political geography also hindered the expansion of Southwest Airlines. The carrier was limited in the markets it could serve from its home airport of Love Field in Dallas, Texas, due to governmental interference. The Civil Aeronautics Board (CAB) forced the cities of Dallas and Fort Worth to combine airport operations at the newly constructed Dallas–Fort Worth International Airport through the Wright Amendment. This limited any airline from using Dallas Love Field to serve any destination that was not in the four states bordering Texas (New Mexico, Oklahoma, Arkansas and Louisiana). Initially this had a small impact on Southwest's network, as the carrier was growing at a slow rate of approximately three cities a year. By the early 1980s, though, the carrier was expanding further and further away from its Texas base and Southwest travellers in the Dallas area were severely hampered by the restrictions placed on Dallas Love Field. Over the years various changes to the amendment were introduced and passed by senators from states outside those allowed service by the Wright Amendment, thereby expanding the number of states served from Dallas Love Field; and by 2014 the amendment will be completely abolished.

Three distinct spatial trends emerge in Southwest's network strategy: a slow growth in markets served based upon distance from the nearest already served Southwest airport; the growth of what are considered secondary airports to serve large metropolitan areas; and an early expansion strategy similar to the Wal-Mart effect.

When Southwest Airlines entered a market in its early expansion stages after deregulation it entered markets generally 500 miles from the nearest already served Southwest airport. The rational for this strategy is quite simple: being a point-to-point carrier at a time when other carriers were establishing and operating hub and spoke networks, Southwest was able to utilize its aircraft more efficiently and offer more frequent daily departures. Spatially, this strategy comes into line with Tobler's first 'law' of geography, that 'Everything is related to everything else, but near things are more related than distant things' (ESRI 2011). This means that cities closer together will have more in common than those farther apart, leading to more possible interaction which, in turn, would lead to the probability of more potential passengers for Southwest. Another geographic model that fits this approach is the Gravity Model: larger cities are more likely to interact than smaller ones and cities closer together have more interactions than those farther apart.

The carrier continued this strategy through the 1980s and 1990s, save 1994 when it acquired Morris Air. This acquisition did not change the geographic strategy of the carrier, as the seven cities added to the network were within the distance from previously served Southwest destinations, as described above. As the carrier grew, the expansion strategy stayed roughly the same, but the addition of longer range aircraft has allowed the carrier to connect cities with direct service that it previously did not have the capability to run. There is a rarely mentioned benefit to Southwest's frequent short haul service: the ability to move passengers through a number of different airports in a hub-like fashion. For example, a passenger flying from Denver to Tampa can be routed through a number of different cities (Houston, Austin, New Orleans, Chicago, and Nashville, to name a few), with the passenger making only one change of aircraft, much like a traditional hub and spoke system. The advantage for Southwest is that the carrier benefits from being able to offer a large number of city pairs without the large costs associated with focusing a large number of resources in a relatively small number of hub airports.

The Wal-Mart effect is the effect a business has in its hinterland, and an increase in its field of influence, that is, an increase in the range of a particular good or service. Southwest Airlines is able not only to introduce the Southwest Effect in the airport in which it offers service but it can also influence the traffic and pricing at airports and on routes from neighboring airports that it does not directly serve. Initially, this effect was the result of Southwest's strategy of basing operations at secondary airports in multi-airport regions.

There are two spatial designations of secondary airports: older airports located closer to traditional downtown areas; and airports located on the fringes of large metropolitan areas and able to serve a wide geographic area, including a number of different cities. Southwest's expansion strategy took advantage of both types of airports; and other LCCs followed suit as they expanded their networks.

Southwest took advantage of older airports, closer to downtown, when it began its operations in the 1970s. Both Dallas Love and Houston Hobby were part of the Texas Triangle mentioned earlier in this chapter. Both airports at the time of entry were slated to be shut down or become afterthoughts, as both cities forced airlines to focus operations on newer airports located farther away from both downtowns. As is mentioned earlier, Southwest's refusal to shift operations to Dallas–Fort Worth International Airport resulted in the passing of the Wright Amendment limiting the destinations Southwest could serve from the airport. Houston Hobby has no such restrictions and, despite the growth of the larger, newer George Bush Intercontinental Airport, has flourished as a focus airport for Southwest and has attracted other LCCs such as Frontier and Jet Blue. In addition to these two airports' closer proximity to downtown Dallas and Houston respectively, another factor emerges as a reason for Southwest choosing to operate out of these smaller, older airports: fortress hubs. Dallas–Fort Worth International Airport is the primary hub for American Airlines and at the time the airport opened, for Braniff International as well. Houston's George Bush Intercontinental Airport is the primary hub for Continental Airlines. Instead of fighting the dominance (measured a number of ways, including gate access, number of flights and pricing power (see Borenstein 1989, Dresner, Windle and Yao 2002) of each of these carriers at their respective hubs, Southwest made the decision to concentrate operations away from the potential operational bottlenecks that fortress hubs can become.

Southwest practised this secondary airport strategy when it expanded into the Chicago market. Instead of basing operations at the larger Chicago O'Hare International Airport, the carrier chose Chicago Midway, which was located closer to the downtown area. Like Dallas/Fort Worth International Airport and George Bush Intercontinental Airport, O'Hare International Airport was a hub airport, but for two, not just for one, carrier, United Airlines and American Airlines. Add to this the fact that operations at the airport can be delayed at times by both weather and congestion and it is easy to see why the location of Chicago Midway was perfect for Southwest's expansion into the Chicago market.

A forgotten aspect of which airports an airline chooses to operate from is the relative location of the airport in question. This is important to both the origin and the destination of a journey, where the customers are starting their journey in a geographic area and where their final destination is within an airport's hinterland. Southwest's focus on serving smaller, more centrally located airports allowed it to promote its service as being closer and therefore more easily accessible to one of the major attractive destinations of a metropolitan area, its downtown. Another way in which Southwest is able to serve a geographically significant area is through its practice of serving secondary airports in a larger geographic area.

Southwest chose to enter a number of large markets, not through the largest airport in the market but instead through secondary airports that were not even in the largest city in the market. Examples of this practice include using Providence, Rhode Island, and Manchester, New Hampshire, as a gateway to the Boston region; serving Fort Lauderdale, Florida, as a gateway to Miami; and using Baltimore/Washington DC International Airport as a gateway to the nation's capital. For a time the carrier dropped services out of San Francisco International and focused on serving San Jose and Oakland as gateways into the Bay Area. Despite the lower costs of using these secondary airports, and the reduced congestion, which enable the carrier to have quick turnaround times and therefore utilize its equipment more efficiently, the question becomes, why does this make sense spatially? There are two reasons why this makes sense spatially: relative location of the airport in question to an origin or a location; and accessibility (Cidell 2006).

The relative location of all four of the airports used as gateways to larger cities shows not only that they are situated to serve the large city but also that a large majority of the population on a larger regional scale is actually closer to these airports than to the 'flagship' airport of the large city. In the cases of Boston, Miami, and Washington DC the largest airport in the region is on the geographic periphery of the larger metropolitan area. Boston Logan is on the far eastern side of Boston, with no population to draw from on the east because of Boston Bay on its east flank. Miami International in more centrally located within Miami, but, on a larger scale of Southeast Florida, is located in the southern third, while Fort Lauderdale International Airport is centrally located to serve Fort Lauderdale, Miami and West Palm Beach. Washington Dulles is in the southwest portion of the metropolitan region and by basing its operations at Baltimore/Washington International (BWI), Southwest became much more accessible not only to Baltimore passengers but also to those in Washington DC, Delaware, western Maryland and southwestern Pennsylvania. Even though Washington's Reagan National is the closest to Washington DC, flight restrictions at the airport limit markets potentially served from that airport.

Accessibility both to and from the airport is another important bonus these airports have over their larger counterparts. As has been mentioned previously, Boston Logan in situated next to Boston Harbor and the only way to access it historically was to travel through downtown Boston. The location is quite good for downtown accessibility, but for passengers who live in the western portion of Boston and surrounding suburbs, Providence and Manchester proved as accessible if not more so than Boston Logan. Southwest has begun services to Boston Logan to six primarily business destinations. This is still fewer than the number of markets Southwest serves out of both Providence and Manchester. Airport accessibility is also an issue at Washington Dulles when compared to BWI. In this particular situation ease of access to BWI is increased by its public transit connections by both local and regional rail services in the terminal and on the airport property in the form of Amtrak. Washington Dulles only offers a bus service to the nearest rail station 14 miles away.

In all four instances Southwest Airlines strategy in choosing to serve secondary airports to gain access to larger metropolitan areas not only positioned the carrier geographically but also spread the 'Southwest effect' to the area as a whole (Vowles 2001, Pitfield 2008, Tierney and Kuby 2008).

Throughout this geographic examination of Southwest Airlines it has been implied that Southwest's growth and success is partially attributed to its following the point-to-point operating system instead of the more widely used hub and spoke operating system. While it is true that Southwest pursues the point-to-point system, it can also be argued that it operates a modified hub and spoke system too. An examination of the carrier's route map shows that certain cities are focal points of the carrier's operations. These focus airports include Las Vegas, Phoenix and Denver in the west, Houston Hobby in the south, Chicago Midway in the Midwest and Baltimore/Washington on the east coast. In the traditional hub and spoke system, an airline will bring in large banks of flights at certain times of the day, have passengers change flights, and then have a large bank of flights depart the hub airport to their final destinations. The difference between that system and the Southwest variant is that Southwest brings in a steady stream of flights during the day and quickly turns around the aircraft, not necessarily waiting for other flights to arrive and for passengers to transfer.

Within the contiguous United States there is not one major airport that does not have services provided by at least one carrier designated as a LCC. The LCC market domestically has matured dramatically over the past 30 plus years. Along with the strategies and the success of LCCs in the United States, the geography of these carriers has changed dramatically over this time. Further, they employ some of the geographic lessons learned from Southwest, including using secondary airports to help their networks grow. This changing geography now leads in two directions. The first is the expansion of LCCs into international markets; while the second sees Allegiant finding a viable niche in providing service from smaller markets to regional leisure destinations. The majority of LCCs in the United States are really an evolution of air travel in the country, and as they mature they begin to look more and more like the majors from a geographic point of view.

Conclusion

The air transport markets and LCCs in both the United States and Canada are among the most mature and sophisticated in the world. While many LCCs have emerged and, subsequently, ceased operations, Southwest Airlines and WestJet are among the most successful airlines in their respective countries. A combination of eased governmental regulations on the industry concerning new entrants and market entry requirements; vast geographies; economies where air travel is both a necessity and a desired commodity for movement and economic growth; and markets geared towards competition has allowed LCCs to flourish domestically and, more recently, internationally. The lessons of the successes and failures of LCCs in this region have guided LCCs in other regions of the world.

References

Barrett, S.D. (2004). How do the demands for airport services differ between full service carriers and low-cost carriers? *Journal of Air Transport Management*, 10(1). Proceedings of the Hamburg Aviation Conference 2003, January 2004, 33–9.

Borenstein, S. (1989). Hubs and high fares: Dominance and market power in the US airline industry. *RAND Journal of Economics*, 20(3), 344–65.

CBC News (1012). WestJet to use Bombardier planes for regional carrier. Available at: http://www.cbc.ca/news/business/story/2012/05/01/westjet-bombardier.html [accessed: 6 July 2012].

Cidell, J. (2006). The regionalization of air travel in central New England. *Journal of Transport Geography*, 14, 23–34.

Dresner, M., Windle, R. and Yao, Y. (2002). The impact of hub dominance and airport access on entry in the US airline industry, in Wang, K.C.P., Xiao, G., Nie, L. and Yang, H. (eds), *Traffic and Transportation Studies: Proceedings of the ICTTS 2002*, Reston, Virginia, 1430–37.

ESRI (2011). *Geography Dictionary, Tobler's First Law of Geography*. Redlands, CA: ESRI Press. Available at: http://support.esri.com/en/knowledgebase/GIS Dictionary/term/Tobler's%20First%20Law%20of%20Geography [accessed: 23 July 2011].

Francis, G., Humphreys, I., Ison, S. and Aicken, M. (2006). Where next for low cost airlines? A spatial and temporal comparative study. *Journal of Transport Geography*, 14(2), March, 83–94.

Goetz, A. and Sutton, C. (1997). The geography of deregulation in the US airline industry. *Annals of the Association of American Geographers*, 87(2), 238–63.

Graham, B. and Vowles, T.M. (2006). Carriers within carriers: A strategic response to low-cost airline competition. *Transport Reviews*, 26(1), 105–26.

Graham, M. (2009). Different models in different spaces or liberalized optimizations? Competitive strategies among low-cost carriers. *Journal of Transport Geography*, 17(4), 306–16.

Howell, D.W., Ellison, R.A., Bateman Ellison, M. and Wright, D. (2003). *Passport: An Introduction to the Tourism Industry* (3rd Canadian ed.). Toronto, Canada: Nelson.

Morrell, P. (2005). Airlines within airlines: An analysis of US network airline responses to low cost carriers. *Journal of Air Transport Management*, 11(5), September, 303–12.

Morrison, S. and Winston, C. (1995). *The Evolution of the Airline Industry*. Washington, DC: Brookings Institution Press.

Nickerson, N.P. and Kerr, P. (2001). *Snaphots: An Introduction to Tourism* (2nd Canadian ed.). Toronto, Canada: Prentice-Hall.

Pitfield, D.E. (2008). The Southwest effect: A time-series analysis on passengers carried by selected routes and a market share comparison. *Journal of Air Transport Management*, 14(3), 113–22.

Putzger, I. (2012). The states men. *Low Cost & Regional Airline Business*, 7(1), 10–12.

RITA BTS (2011). *Research and Innovative Technology Administration*. Available at: http://www.transtats.bts.gov/carriers.asp [accessed: 27 July 2011].

RITA BTS (2012). *Research and Innovative Technology Administration*. Available at: http://www.transtats.bts.gov/ [accessed: 25 August 2012].

Tierney, S. and Kuby, M. (2008). Airline and airport choice by passengers in multi-airport regions: The effect of Southwest Airlines. *Professional Geographer*, 60(1), 15–32.

UNWTO (2012). *UNWTO Tourism Highlights 2012 Edition*. Madrid: World Tourism Organisation. Available at: https://s3-eu-west-1.amazonaws.com/storageapi/sites/all/files/docpdf/unwtohighlights12enhr_1.pdf [accessed: 25 June 2012].

Vowles, T.M. (2001). The 'Southwest Effect' in multi-airport regions. *Journal of Air Transportation Management*, 7(4), 251–8.

WestJet (2011). *WestJet: 2011 Annual Information Form: March 14, 2012*. Available at: http://www.westjet.com/pdf/investorMedia/financialReports/WestJet2011AIF.pdf [accessed: 12 June 2012].

Windle, R. and Dresner, M. (1999). Competitive responses to low cost carrier entry. *Transportation Research E: The Logistics and Transportation Review*, 35, 59–75.

Windle, R., Lin, J. and Dresner, M. (1996). The impact of low-cost carriers on airport and route competition. *Journal of Transport Economics and Policy*, 30(3), 309–28.

Chapter 5
Low Cost Carriers in South America

Gui Lohmann and Gustavo Lipovich

Introduction

The South American subcontinent, which does not include the Caribbean, Mexico and other countries in Central America, has not been traditionally recognized as one of the most dynamic air transport regions in the world. In 2010, the region's participation in the worldwide market share of available seats offered represented only 6.4 per cent. However, the 2000s brought some political, regulatory and market strategy changes into this region. As a consequence, the total number of available seats in the region increased 40.5 per cent between 2001 (169 million available seats) and 2010 (237 million available seats) (OAG 2010). Those changes created an environment in which LCCs could flourish. As a result, while the number of available seats offered by LCCs grew by almost 375 per cent worldwide between 2001 and 2010, in South America it increased almost ten times in the same period. Data on low cost carrier (LCC) seat capacity shows that the LCC sector in South America jumped from a worldwide participation of 2.74 per cent in 2001 to nearly 7 per cent in 2010 (see Table 5.1). Detailed analysis shows that since 2003, Brazil has completely dominated almost the entire LCC market share in South America, which corresponds to an average market share of 98.11 per cent in the period 2003–2010.

Table 5.1 Yearly seat capacity offered by low cost carriers (2001–2010)

Year	Worldwide (W)	South America (SA)	SA/W (%)	Brazil (Br)	Br/SA (%)
2001	232,706,647	6,386,427	2.74%	2,542,204	39.81%
2002	267,606,232	10,484,795	3.92%	7,918,063	75.52%
2003	323,165,425	12,040,047	3.73%	11,586,215	96.23%
2004	410,563,914	13,348,761	3.25%	13,245,621	99.23%
2005	474,080,924	15,722,430	3.32%	15,565,218	99.00%
2006	549,996,621	23,528,121	4.28%	23,408,165	99.49%
2007	674,700,115	30,675,963	4.55%	30,097,623	98.11%
2008	752,607,102	35,317,179	4.69%	34,452,519	97.55%
2009	770,280,199	49,565,719	6.43%	48,386,208	97.62%
2010	873,975,749	60,968,101	6.98%	59,551,127	97.68%

Source: OAG 2010.

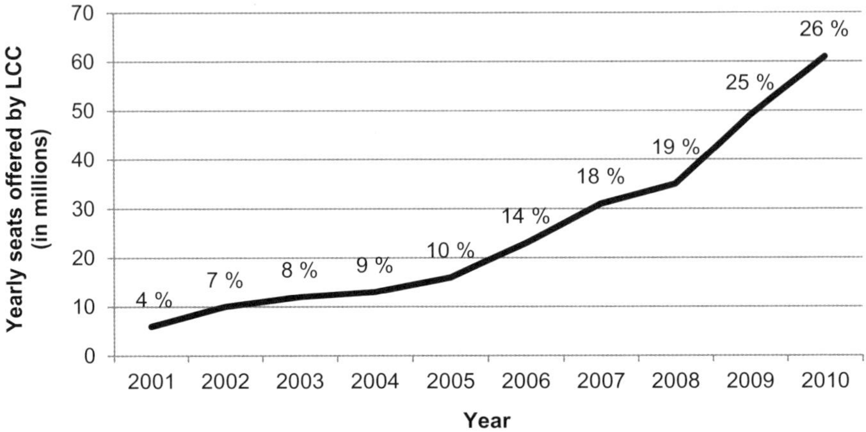

Figure 5.1 Yearly share of low cost carrier seat capacity in comparison to the total South American air transport market

Source: After OAG 2010.

An analysis of the air transport market in South America during the period 2001–2010, LCCs' seat capacity increased from 4 per cent in 2001 to 26 per cent in 2010 (see Figure 5.1). Considering the significant growth of the LCC sector in South America it is worth analysing the LCC phenomenon in the region and, in particular, the case of Brazil. Hence, this chapter provides an overview of the external macroeconomic changes in South America that have contributed to the development of LCCs in the region, followed by a more detailed analysis of Brazil, the region's predominant LCC market. The timeframe for both analyses is the period from 2001 to 2010. Conclusions are then drawn, which present some lessons from the South American LCC sector.

Overview of Low Cost Carriers in South America

In order to understand the proliferation of LCCs in South America and the dominance of Brazil, it is crucial to analyse the aviation and non-aviation factors that have changed in the region during the 2000s.

Non-aviation Factors

In comparison to other countries in South America, Brazil has some unique characteristics that fostered the flourishing of LCCs. In terms of its landmass, population size and gross domestic product (GDP), Brazil accounts for approximately half of each of these three factors in South America, favouring an

Table 5.2 Area, population and GDP of major countries in South America

	Area (sq km)	% Total area	Population*	% Total population	GDP (US$ billions)**	% Total GDP	GDP per capita (US$)
Argentina	2,780,400	15.68	41,769,726	10.45	596.00	13.74	14,269
Bolivia	1,098,581	6.20	10,118,683	2.53	47.88	1.10	4,732
Brazil	8,514,877	48.02	203,429,773	50.88	2,172.00	50.09	10,677
Chile	756,102	4.26	16,888,760	4.22	257.90	5.95	15,271
Colombia	1,138,910	6.42	44,725,543	11.19	435.40	10.04	9,735
Ecuador	283,561	1.60	15,007,343	3.75	115.00	2.65	7,663
Guyana	214,969	1.21	744,768	0.19	5.38	0.12	7,224
Paraguay	406,752	2.29	6,459,058	1.62	33.31	0.77	5,157
Peru	1,285,216	7.25	29,248,943	7.32	275.70	6.36	9,426
Suriname	163,820	0.92	491,989	0.12	4.71	0.11	9,573
Uruguay	176,215	0.99	3,308,535	0.83	47.99	1.11	14,505
Venezuela	912,050	5.14	27,635,743	6.91	345.20	7.96	8,006
Total	17,731,453	100.00	399,828,864	100.00	4,336.47	100.00	10,845

Note: * Estimated July 2011; ** Estimated 2010.
Source: CIA 2011.

overall need for the development of air transport (see Table 5.2). Although Brazil has the highest GDP in the region, the high level of interpersonal and interregional inequalities existing in the country ensure that it does not have the highest GDP per capita in the region, but rather ranks behind Chile, Uruguay and Argentina (see Table 5.2). For example, according to Forbes (2011), Brazil has the eighth largest number of billionaires in the world, a number much larger than that found in other Latin American countries, and which is even ahead of Japan. On a regional level, in 2008, a quarter of the country's GDP was concentrated in six state capital cities, namely São Paulo (11.8 per cent), Rio de Janeiro (5.1 per cent), Brasília (3.9 per cent), Belo Horizonte (1.4 per cent), Curitiba (1.4 per cent) and Manaus (1.3 per cent). The state of São Paulo alone represents a third of the country's GDP. As a consequence, the wealth of some of its metropolises and some segments of its population creates a high demand for the use of air transport to cover the distance of the continental landmass.

Some countries in South America, such as Guyana, Paraguay, Uruguay and Suriname, have too small a landmass and economy to sustain a large aviation sector (see Table 5.2). In spite of its landmass, Bolivia has the lowest GDP per capita in the region, hence it has fewer opportunities to develop air transport. Argentina, Brazil, Chile, Colombia, Ecuador, Peru and Venezuela present a combination of population, landmass and economic development to support their aviation markets.

Aviation-based Conditions to Support the Proliferation of Low Cost Carriers in South America

In order to implement their business models and thus proliferate, LCCs require a macro aviation environment apart from non-aviation conditions. In the following paragraphs, some main aviation conditions that influence the support for LCCs are presented and contextualized in the case of South America.

Liberalization of airline markets is the main aviation condition. According to Francis et al. (2006: 90), 'liberalization of domestic markets [is] closely followed by low cost airline activity'. In this sense, there are two crucial aspects to the liberalization of airline markets: the freedom to set fares; and unrestricted opportunities for new flights, frequencies, destinations and equipment. Prior to deregulation there were too many impediments to flexible airline management to operate for the best economic results. Francis et al. (2006: 92) argue that 'without freedom to enter routes and to charge what fares an operator wishes, the low cost model is impeded in its development'. Thus, it is pertinent to affirm that after several years of normative transformation, nearly all of the South American airline markets were liberalized. Chile, Peru and Colombia liberalized their domestic markets during the 1990s (see Villena, Harrison and Villena 2008), while in Brazil, liberalization of the domestic airline market was fully achieved in 2005 (Bettini and Oliveira 2008b). The exceptions are Argentina, Bolivia, Ecuador and Venezuela, which experienced liberalization in the 1990s but partial re-regulation during the 2000s (Lipovich 2010).

Internationally, there are two major airline market liberalization agreements in South America. The Commission of the Andean Community of Nations (CAN) signed an open skies agreement in 1991 (Goh 2001 and Roessing Neto 2007). This was originally signed by Bolivia, Colombia, Ecuador, Peru and Venezuela, although the last has excluded itself from the agreement by formally withdrawing from the CAN. Moreover, six countries signed the Fortaleza Agreement in 1996: Argentina, Bolivia, Brazil, Chile, Paraguay and Uruguay; and Peru joined in 1999. This agreement liberalized regional international routes that were not included in bilateral agreements, although some of the latter were made flexible enough or fully liberalized (Lipovich 2009). Apart from these international agreements within South America, Chile, Peru, Paraguay, Uruguay, Colombia and Brazil signed open skies agreements with the United States (Lipovich 2010). These bilateral agreements signed with the United States abolished several traffic restrictions by establishing unlimited market access, multiple designations, no frequency or capacity control and free pricing. Although liberalization is not yet complete, it is possible to argue that there is a very high degree of deregulation in the South American airline market.

In relation to the existence of strong traffic flows, there are very high-demand routes that link densely populated cities, both among themselves and with some South American tourist destinations that have a high national and international influx. The increase in air demand takes place in the context of the sustained economic growth that has occurred since 2002.

However, at popular destinations, it is possible to find slot-constrained hub airports, although almost all large cities have under-utilized airports that are currently used for general and military aviation but that have the potential to change their future use to commercial aviation. This is a significant element of the attraction of LCCs, as supported by Lawton (2000, 2003) and Doganis (2006). Also, there are many commercial airports that have spare capacity. In this sense, it is relevant that 'for secondary airports the deregulated market brought the opportunity to develop business with the new market entrants rather than remain underutilized' (Barret 2004: 91) This was the case of Azul, which is one of the most recent LCCs to be established in Brazil, as it had issues entering the saturated airports in the city of São Paulo and opted to use Campinas (90 kilometres from São Paulo) and its under-utilized airport.

Recently, commercially oriented airport management has grown in South America. According to Lawton (2000), Alamdari and Fagan (2005) and Dobruszkes (2006), this encourages the development of LCCs due to their objective of occupying unused slots and increasing airport revenues. Though it is not a mass practice, all major international airports in South America – for example, Buenos Aires (Ezeiza), Santiago de Chile, São Paulo (Guarulhos), Lima and Bogota – have adopted this business strategy in one way or another, regardless of whether they are managed by public or private entities.

Inversely, the expenditures of public local subsidies to attract new commercial air services is not very common in South America. These kinds of subsidies have encouraged LCC operations in different parts of Europe and the USA (see Alamdari and Fagan 2005, Barret 2004, Dobruszkes 2006 and Francis et al. 2006). However, because countries in South America usually have other, more relevant social demands, such as health, education, housing and law and order, investment in air transport is not seen as a priority.

Another macro condition to stimulate the spread of LCCs that could not be found in South America is the minimal unionization among employees. In this regard, when unionization is high, there is little job flexibility, the presence of collective agreements is very common, and the relationships between workers and airlines are not generally established on an individual basis. Thus, in a unionized environment, outsourcing possibilities are minimized and labour productivity is affected because workers for LCCs tend to have higher workloads and usually are paid less than their fellow workers at other airlines (Dobruszkes 2006).

The last of the external conditions that usually was not found in South America during the 2000s is the presence of a financial crisis in airline markets. According to Lipovich (2005) and Doganis (2006), this kind of crisis, which can respond to macroeconomic, social or natural factors, influences and pressures the airlines to lower their costs. In general, crises encourage the growth of existing LCCs and stimulate the emergence of new ones in this context.

In summary, the presence of the most significant external conditions that support the proliferation of LCCs is not completely present in South America, although several of these conditions are favourable. Table 5.3 reflects a subjective estimation of these external conditions as an average for the whole region.

Table 5.3 Presence of macro-conditions to facilitate the flourishing of low cost carriers in South America

Macro-conditions	Presence
Liberalization of airline markets	Yes (high)
Strong traffic flows	Yes (in some cases)
Existence of under-utilized airports	Yes (many)
Airport marketing	Yes (a few)
Local public subsidies	No (low)
Minimal unionization	No (low)
Financial crisis in airline markets	No (low)

Overview of Low Cost Carriers in South America

The specialized literature generally recognizes the existence of different LCC business models, which can be classified according to their origins or characteristics. Alamdari and Fagan (2005) applied a method to compare different models of LCCs in which some strategies are weighted by how similar they are to the original blueprint model developed by Southwest Airlines. In fact, the basis of the comparison is to check the implementation of 17 main commercial strategies, including the following (see also Chapter 1):

- *Networks and ticketing:* no connecting, through fares, no interlining, simple point-to-point sectors with only one-way fares;
- *Services:* single class configuration, in-flight frills, seat assignments, frequent flyer programme, designated cargo;
- *Distribution:* travel agents, online booking, codeshares;
- *Operational features:* fleet commonality, average aircraft utilization, average stage length, airports served.

While the above list is not exhaustive, the listed items provide a comprehensive comparative analysis. The model proposed by Alamdari and Fagan (2005: 381) involves a calculation based on assigning a score for each of the 17 commercial strategies: 'a score of two is assigned to a feature that is identical to the original model; a similar feature is scored one; a feature that is completely different is scored zero'. Inevitably, airlines will have a score ranging from 0 to 34. The index is calculated out of 100, the value received for a carrier that fully adheres to the original model.

Due to the rise of airline hybrid business models, currently few airlines achieve an index that is close to 100. For example, Figure 5.2 presents the indexes for a number of airlines in South America, and no value exceeds 80. Unlike the calculation performed by Alamdari and Fagan (2005), Figure 5.2 also provides the index for some Full Service Network Carriers (FSNCs) in order to verify the differences between LCCs and FSNCs in terms of cost reduction strategies implemented.

To better identify and analyse the LCCs in South America, Table 5.4 lists those airlines presented in Figure 5.2 with an index higher than 25, in addition to those airlines that self-identified as LCCs on their websites. With the exception of Aires Airlines, all other airlines listed in Table 5.4 initiated their operations as passenger airlines in the 2000s, showing that, as previously discussed, this period has been favourable in South America for the establishment of LCCs. Table 5.4 provides an opportunity for comparison between the Brazilian and the remaining South American LCCs. Brazil is the country served with the highest number of LCCs (four in total). Each of the Brazilian LCCs serves at least 20 airports and makes use of a proportionally larger fleet of aeroplanes, comprising mainly Airbus and Boeing jets. This is consistent with the large size of the LCC market in Brazil, as presented in Table 5.1. Non-Brazilian South American LCCs serve fewer airports and have either smaller fleets of large jets (e.g. Andes Líneas Aéreas, BoA and Peruvian Airlines) or a fleet consisting of

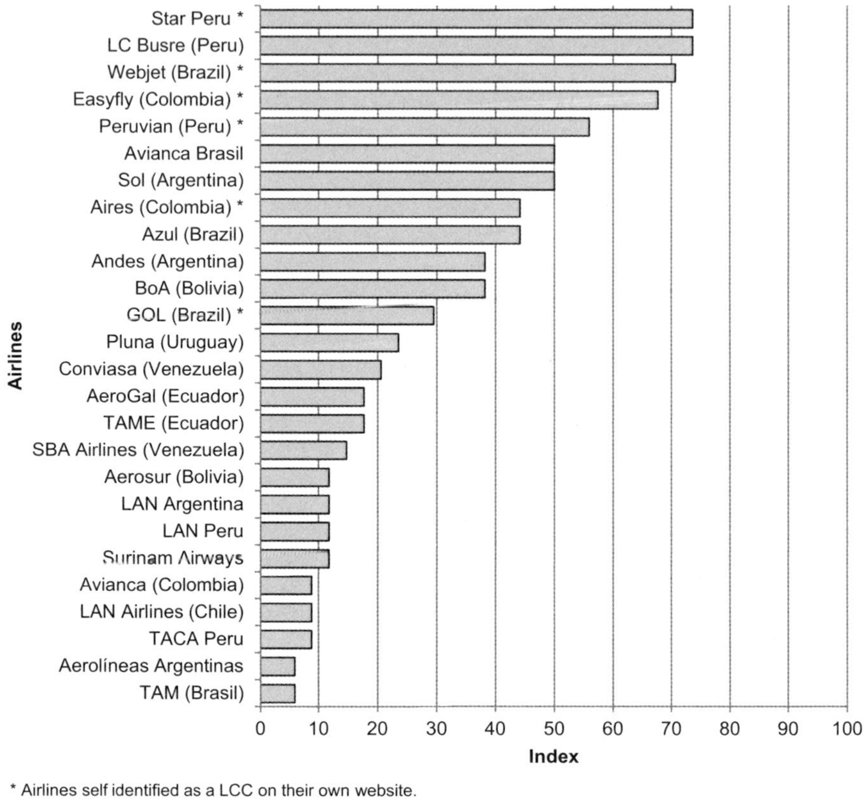

Figure 5.2 Index of adherence of South American airlines to the original low cost carrier model

Source: Proposed by Alamdari and Fagan (2005). The authors elaborated data from April 2011 airline information.

small regional jets (e.g. Sol Líneas Aéreas, EasyFly, LC Busre and Star Peru). The only exception is Aires, in Colombia, with a mixed fleet of B737s and regional turboprop aircraft.

Of the countries presented in Table 5.4, it is worth mentioning that Argentina, Chile and Uruguay represent the three countries with the highest GDP per capita in South America (see Table 5.2). Argentina could have a larger number of LCCs if it was not for the lack of a deregulated market to encourage the entrance of competitive LCCs. Chile is very much dominated by the LAN group, and in terms of population and area, Uruguay has a very small market to justify a LCC operation.

Table 5.4 South American airlines with an index of adherence to the original low cost carrier model higher than 25

Country	Airline	Year started operating	Airports served	Fleet (aircraft passenger capacity)
Argentina	Andes Líneas Aéreas	2006	11	7 MD-80 (165) 2 CRJ-900 (90)
	Sol Líneas Aéreas	2006	11	6 Saab 340 (34)
Bolivia	BoA	2007	7	6 B737-300 (138)
Brazil	Avianca Brazil	2002	24	14 Fokker 100 (100) 3 A318 (120) 3 A319 (132)
	Azul	2008	40	8 ATR-72 (68) 10 E190 (106) 22 E195 (118)
	GOL	2001	74	42 B737-700 (144) 73 B737-800 (178-189) 1 B767-200 (221) 3 B767-300 (218-258)
	Webjet	2005	20	24 B737-300 (148)
Colombia	Aires	1981	24	9 B737-700 (149) 11 Dash 8Q-200 (37) 4 Dash 8Q-400 (78)
	EasyFly	2007	15	12 BAe-Jetstream 41 (30)
Peru	LC Busre	2001	7	6 Fairchild Metro III (19)
	Peruvian Airline	2009	4	4 B737-200ADV (120)
	Star Peru	2004	15	9 BAe 146 (77-112)

Sources: Airline websites.

The Low Cost Carrier Experience in Brazil: 2001–2011

Aviation Profile in Brazil

Although there are no precise data regarding actual domestic tourism transport demand in Brazil, a survey commissioned in 2006 by the Brazilian Ministry of Tourism stated that domestic tourism trips are made predominantly by road transport, with private cars (45.7 per cent), coaches (25.5 per cent) and charter coaches (7.9 per cent) accounting for eight out of ten trips, and air transport representing 12.1 per cent of domestic travel. In Brazil, rail and water transport for regional and interstate transport is almost nonexistent.

Until 2001, the Brazilian airline industry was dominated by the 'Big Four' – VARIG, VASP, TAM, and Transbrasil – when GOL Linhas Aéreas (Gol)

Figure 5.3 SOL Líneas Aéreas uses a fleet of Saab 340 regional turboprop aircraft

Source: Michael Lück.

introduced its operation as the first low cost carrier in the country (Costa, Lohmann and Oliveira 2010, Oliveira 2008). The following years were characterized by strong growth but also by competition in the aviation industry, which bankrupted inefficient airlines such as Transbrasil and VASP in 2001 and 2002, respectively (Huse and Oliveira 2010). In March 2007, Gol purchased VARIG, and since that time, Gol and TAM together have held approximately 94 per cent of the domestic market share (IATA 2009).

In 2003, due to excess in capacity and predatory competition in the market, the then Department of Civil Aviation (DAC), currently ANAC (Civil Aviation National Agency), partially re-regulated the aviation market by temporarily banning new aircraft imports, controlling price competition and disallowing strategic movements towards increased market concentration (Bettini and Oliveira 2008a and Oliveira 2006). Nevertheless, air traffic, particularly between key airports located in the country's major metropolises, increased rapidly in the following years and opened new challenges for the aviation sector's lagging infrastructure (Costa, Lohmann and Oliveira 2010). The unequal development of infrastructure and resource allocation led to the Brazilian aviation crisis in the years 2006 and 2007, when there occurred a series of flight delays and cancellations, strikes, incidents and the two worst deadliest accidents in Brazilian aviation history. Although the Brazilian government emphasized updating and expanding airport infrastructure and safety, a report by McKinsey & Company (2010) concluded that seven of Brazil's 20 principal airports were struggling with overcrowded passenger areas and plane gates, which frequently led to flight delays or cancellations.

Development of Low Cost Carriers in Brazil

As in many other parts of the world, air transport in Brazil went through a deregulation process which aimed to improve competition among airlines. The Air Transport Liberalization Policy established in the 1990s created an environment in which new airlines could access air transport markets, increasing their competitiveness. As a result, the 2000s brought some radical changes in Brazil's domestic aviation sector, with the January 2001 launch of Gol (see details below). During this period, the dismissal of three inefficient legacy carriers took place, with VARIG re-emerging as the 'New VARIG' in 2006, and finally bought by Gol in 2007 (for more details, see Espírito Santo Jr 2008).

In 2003, after a certain period of economic freedom, regulation was imposed again due to what was considered to be an 'excess in capacity' and 'predatory competition' in the market (Bettini and Oliveira 2008b). Nevertheless, there are clear signs that the Liberalization Policy brought many benefits, such as lower airfares, higher operational efficiency and competitiveness through market expansion. However, the same economic liberalism was not seen in the sector's lagging infrastructure, including airports and air traffic control, which stayed under government supervision. These factors combined built a peculiar network design that was highly concentrated in a few airports in São Paulo and Brasília (see Figure 5.4), and which contributed to two major fatal accidents in 2006–2007, which created chaos in the air transport system with delays and flight cancellations over several months (Costa, Lohmann and Oliveira 2010).

Towards the end of the 2000s, other airlines operating in the LCC segment entered this market, namely Webjet and Azul. In 2005, Webjet started operation as a charter airline but changed its business model in 2009 to that of an LCC offering scheduled flights. In 2011, Gol bought Webjet. In 2008, Azul was launched by David Neeleman, former CEO of JetBlue in the US. In its first three years Azul has operated predominantly with Embraer regional jets. Azul is the first LCC in Brazil to establish its main hub at a secondary airport, in this case, Campinas airport. Its main advantage is that in contrast to the congested airports of Congonhas and Guarulhos, Campinas is a large, under-utilized freight-focused airport.

As a result of the changes in the air transport market in Brazil during the 2000s, TAM and Gol were consolidated as the two main domestic airlines, accounting for over 80 per cent of domestic air transport capacity by 2010 (see Figure 5.5). With the dismissal of legacy carriers such as Transbrasil and VASP and the establishment of new LCCs (i.e. Azul and Webjet), the current remaining market share belongs to regional airlines. Figure 5.5 illustrates these changes between the years 2002 and 2010 from the capacity point of view. Figure 5.6 compares the participation of LCCs in Brazil in terms of frequency of flights and seat capacity. By 2010, LCCs

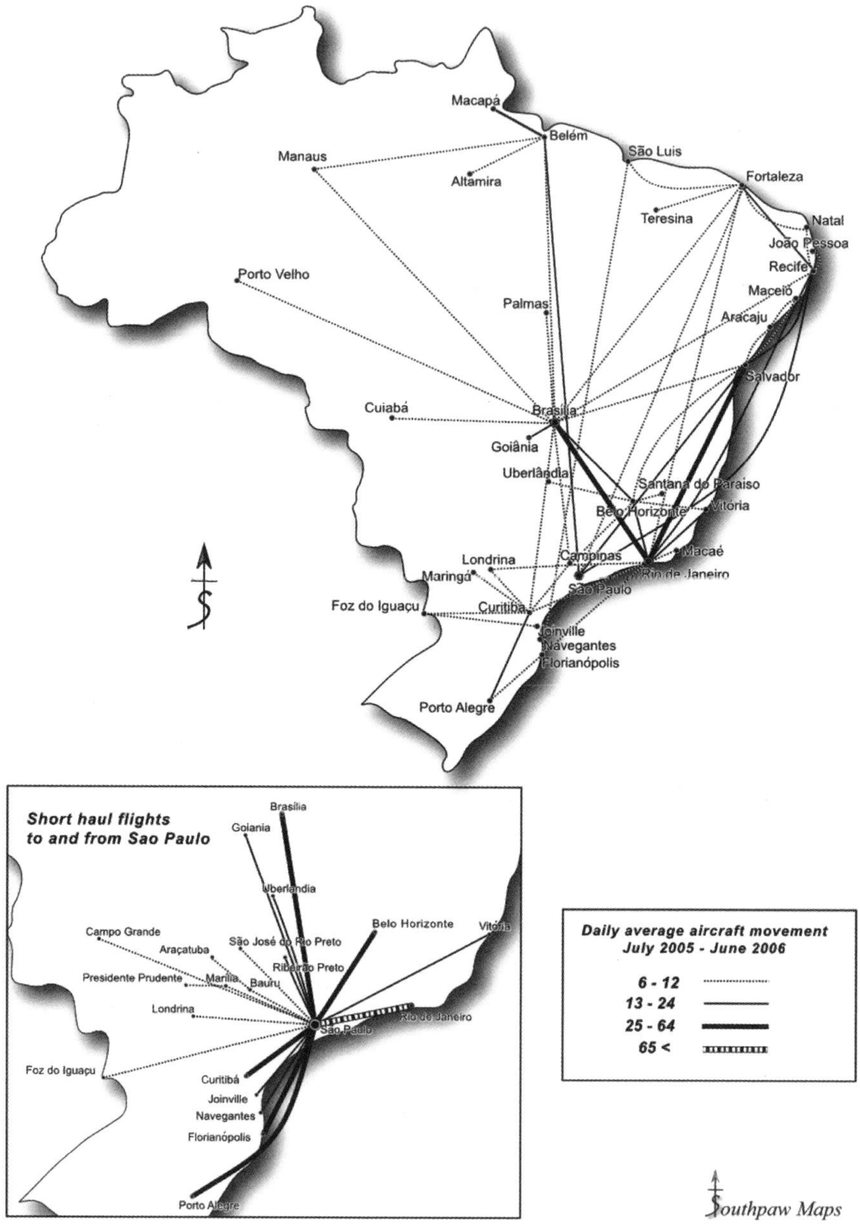

Figure 5.4 **Aircraft movements to top airports in Brazil during the period July 2005–June 2006**

Note: Only routes with six or more average daily flights are presented.

Source: Costa, Lohmann and Oliveira 2010.

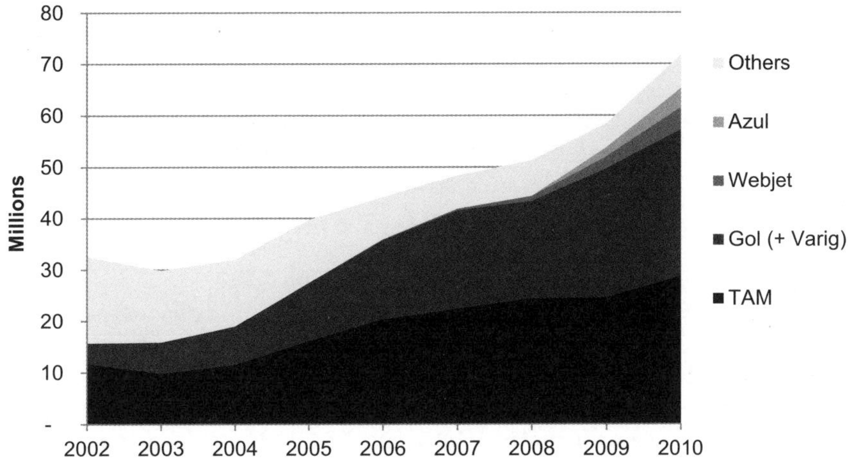

Figure 5.5 Domestic air transport passengers carried by major airlines in Brazil

Source: ANAC 2011.

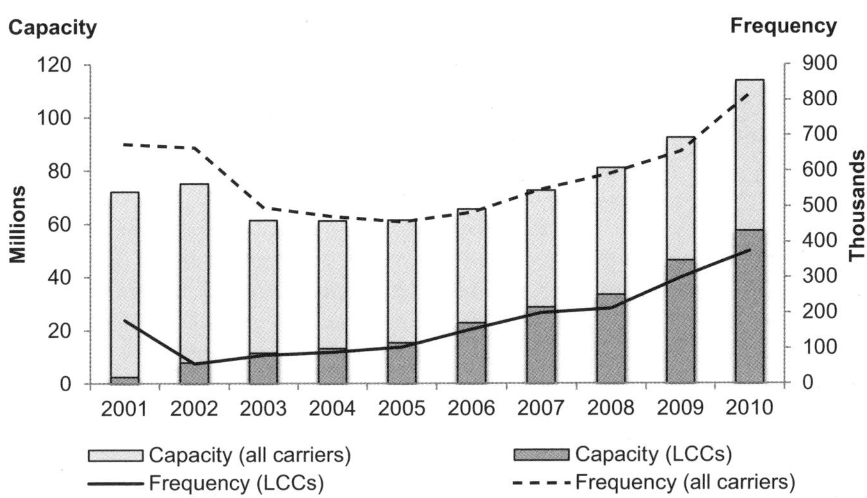

Figure 5.6 Yearly domestic seat capacity and frequency for low cost carriers and all airlines in Brazil

Source: OAG 2010.

represented nearly half of the market for both variables and showed an overall sustainable growth since 2003.

Case Study: The Case of GOL Linhas Aéreas

After the removal of regulatory barriers and the liberalization of the Brazilian airline industry in 1992, GOL Linhas Aéreas (Gol) was not only the first Brazilian LCC but was also transformed into the largest LCC in South America (see comparison in Table 5.5). Gol started operations in January 2001 (Oliveira 2006). By using only a few LCC strategies such as cutting catering, using a standardized fleet and putting strong emphasis on e-ticketing, Gol's unit costs and yields were one-third lower than those of its competitors (Oliveira 2008).

Consequently, after its second year of operations, Gol was the only profitable major airline in Brazil at that stage, with an operating margin of 5.8 per cent of total revenues (DCA 2002). With a share of 12 per cent of the domestic market in 2002, Gol became one of the strongest competitors of the legacy carriers, which at that stage also included VARIG, VASP and TAM (Oliveira 2008). In the following two years, Gol's expansion was influenced by two factors (Bettini and Oliveira 2008a, Oliveira 2005): First, in March 2003, Gol successfully traded 20 per cent of its equity share (worth US$26 million) to American International Group (AIG), which enabled the acquisition of additional aircraft. Second, due to overcapacity and over-competition in the market, the former Department of Civil Aviation (DAC, currently ANAC) banned new aircraft imports, controlled price competition and disallowed strategic movements towards increased market concentration.

Despite these market regulations, Gol's domestic market share rose to 30 per cent in 2005 after Transbrasil and VASP stopped their operations (Huse and Oliveira 2010). Subsequently, Gol overtook VARIG and placed itself in second place in the Brazilian domestic market behind TAM (de Araújo Júnior, dos Santos, Piers 2007). In March 2007, Gol purchased VARIG and relaunched it with specialization in charter flights, upper-scale service and international destinations (Gol 2011). The acquisition of VARIG enabled Gol's expansion to international destinations, including flights to North America, Europe and Africa using Boeing 767-300s (Bettini and Oliveira 2008a). In 2009, Gol was merged into VRG Linhas Aéreas with two separate operating brands, Gol and VARIG, both of which operated under Gol flight numbers and used Gol's booking system (Gol 2011). In August 2011, VRG Linhas Aéreas purchased Webjet Linhas Aéreas (see the description of Webjet above under 'Development of low cost carriers in Brazil', and *World Airline News* 2011).

When it was launched, Gol operated with only six Boeing 737-700s flying to key destinations in Brazil, including São Paulo, Rio de Janeiro, Belo Horizonte, Florianópolis, Brasília, Porto Alegre and Salvador. Every aircraft thereby operated for approximately 10 to 12 flights a day, and all direct routes were below 1,500 kilometres (Oliveira 2005). By the end of 2002 Gol increased its fleet to 22 aircraft, which were either Boeing 737-700s or Boeing 737-800s, and served a much wider network of 54 direct routes and distances up to 3,000 kilometers (Oliveira 2008). In 2004, Gol started operating in the international market, and

Table 5.5 Overview of Gol's key developments and other major facts pertaining to the domestic aviation market in Brazil, 2001–2011

Year	Event
2001	Gol starts operation with six Boeing 737-700s and 33 direct routes within the Brazilian domestic market. Transbrasil ceases its operations.
2002	Gol increases its fleet to 22 aircraft and its network to 54 direct routes within the Brazilian domestic market. Its Brazilian domestic market share rises to 12%. Airline VASP stops its operations.
2003	Gol trades 20% of its equity share (worth US$26 million) to American International Group (AIG) and acquires additional aircraft. The former Brazilian Department of Civil Aviation (DAC) re-regulates the market due to over-competition and overcapacity.
2004	At the end of the year, Gol launches its first international flight between São Paulo and Buenos Aires. In the following years, it would expand its international network to other countries in Latin America.
2005	Gol gains 30% market share and places itself in second place in the Brazilian domestic market, behind TAM. VARIG declares bankruptcy in June 2005.
2006	A Gol Boeing 737-800 collides with a business jet; none of the passengers and crew members on board survive.
2007	Gol purchases VARIG and relaunches VARIG with specialization on charter flights and flights to international destinations. Gol expands its network to international destinations in North America, Europe and Africa, using Boeing 767-300ERs.
2008	Gol operates in 50 airports and has a network of more than 200 direct routes.
2009	Gol merges into VRG Linhas Aéreas with the two operating brands of Gol and VARIG.
2011	Gol purchases WebJet Linhas Aéreas. Gol serves 61 destinations with 108 aircraft and 860 daily flights.

Sources: Various sources cited in this chapter.

as of October 2011, it flies to a number of countries in South America, as well as Panama and the Caribbean. By 2008, Gol operated at almost 50 airports and expanded its network to more than 200 domestic and international direct routes (Huse and Oliveira 2010). A decade after its launch, Gol served 61 destinations by operating 108 aircraft and 860 daily flights (Gol 2011).

Conclusions – The Low Cost Carrier Sector in South America: 2001–2010

This chapter has provided an overview of LCCs in South America, with an emphasis particularly on the 2000s. Although South America has only a small market share in terms of the worldwide LCC seat capacity (nearly 7 per cent in

2010), the region experienced significant growth in the first decade of the twenty-first century. A more detailed analysis shows that Brazil has been the powerhouse behind the growth of LCCs in South America, currently dominating the South American market with over 97 per cent in terms of seat capacity. This chapter provided an overview of the aviation and non-aviation factors existing in South America that could encourage LCCs to flourish. Some South American countries have the necessary market to support an LCC operation but not the regulatory framework to facilitate the entrance of LCCs, while others are too small in terms of demand, land mass or economic power to foster an LCC. Using the LCC adherence index proposed by Alamdari and Fagan (2005), a list of LCCs and FSNCs was generated, with airlines from Brazil (e.g. Avianca, Webjet, Azul and Gol), Colombia (Easyfly and Aires) and Peru (LC Burse and Peruvian) identified at the top of the list as those with strong LCC business models.

Among those LCCs, Gol deserves special attention, as it has the largest fleet in the region, comprising a mix of B737-700s and B737-800s. All other LCCs in South America either have smaller fleets of B737s or large fleets of regional jets or propeller-driven aircraft. Because Gol is the major LCC in Brazil, the country that dominates the LCC market in South America, this chapter has presented a description of Gol's evolution since its origins in 2001.

Although Gol has followed some of the original blueprint characteristics of LCCs, particularly the characteristics of Southwest, it has not operated out of secondary airports (the first airline to do so in Brazil was Azul). It has initiated international flights to other destinations in Latin America and it has acquired other airlines with different business models such as VRG Linhas Aéreas (the new version of the legacy carrier VARIG) and Webjet (which used to operate as a charter airline). Gol is one of many airlines around the world (e.g. Virgin Australia, Air Asia and Ryanair) that have adjusted their business models to fit opportunities, competition and regulatory environments to their own realities. In spite of Brazil's deficiencies in airport and air service infrastructure, it has been a dynamic environment for new LCCs, including Azul. Further and more detailed studies will be required in this area.

References

Alamdari, F. and Fagan, S. (2005). Impact of the adherence to the original low-cost model on the profitability of low-cost airlines. *Transport Reviews*, 25(3), 377–92.

ANAC (2011). About ANAC. Brasília, Brazil: National Civil Aviation Agency (ANAC). Available at: http://www.anac.gov.br/portal/cgi/cgilua.exe/sys/start.htm?sid=341 [accessed: 18 September 2011].

Barret, S.D. (2004). The sustainability of the Ryanair model. *International Journal of Transport Management*, 2, 89–98.

Bettini, H.F. and Oliveira, A.V.M. (2008a). *Airline Capacity Setting After Re-regulation: The Brazilian Case in the Early 2000s*. São Paulo, Brazil: Center for Studies of Airline Regulation and Competition.

Bettini, H.F and Oliveira, A.V.M. (2008b). Airline capacity setting after re-regulation: The Brazilian case in the early 2000s. *Journal of Air Transport Management*, 14, 289–92.

CIA (2011). *The World Factbook*. Available at: http://www.cia.gov/library/publications/the-world-factbook/index.html [accessed: 29 August 2011].

Costa, T.F.G., Lohmann, G. and Oliveira, A.V.M. (2010). A model to identify airport hubs and their importance to tourism in Brazil. *Research in Transportation Economics*, 6(1), 3–11.

DCA (2002). *Statistical Yearbook of the Department of Civil Aviation*. São Paulo, Brazil: Department of Civil Aviation.

de Araújo Júnior, A.H., dos Santos, I.C. and Pires, C.C. (2007). A comparative study of the relative efficiency of American, European, Asian and South American Airlines, in *Proceedings of 11th Annual World Conference of the Air Transport Research Society*, University of California, 23 June 2007, Berkeley, CA.

Dobruszkes, F. (2006). An analysis of the European low-cost airlines and their networks. *Journal of Transport Geography*, 14, 249–64.

Doganis, R. (2006). *The Airline Business in the 21st Century* (2nd ed.). London: Routledge.

Espírito Santo Jr., R.A. (2008). Brazil, in Graham, A., Papatheodorou, A. and Forsyth, P. (eds), *Aviation and Tourism: Implications for Leisure Travel*. Aldershot, UK: Ashgate, 257–65.

Forbes (2011). Behind Brazil's billionaire boom. Available at: http://www.forbes.com/sites/kerenblankfeld/2011/03/09/behind-brazils-billionaire-boom [accessed: 29 August 2011].

Francis, G., Humphreys, I., Ison, S. and Aicken, M. (2006). Where next for low cost airlines? A spatial and temporal comparative study. *Journal of Transport Geography*, 14, 83–94.

Goh, J. (2001). *The Single Aviation Market of Australia and New Zealand*. London: Cavendish.

Gol (2011). About us. Available at: http://www.voegol.com.br/int/Gol/AboutUs/Paginas/home.aspx [accessed: 20 August 2011].

Huse, C. and Oliveira, A.V.M. (2010). *Does Product Differentiation Soften Price Reactions to Entry? Evidence from the Airline Industry*. São Paulo, Brazil: Center for Studies of Airline Regulation and Competition.

IATA (2009). *The Impact of International Air Service Liberalisation on Brazil*. Montreal, Canada: International Air Transport Authority.

Lawton, T.C. (2000). Flying lessons: Learning from Ryanair's cost reduction culture. *Journal of Air Transportation World Wide*, 5, 89–105.

Lawton, T.C. (2003). Managing proactively in turbulent times: Insights from the low-fare airline business. *Irish Journal of Management*, 24, 173–93.

Lipovich, G. (2005). New trends in scheduled air transport: The crisis of the sector, low-cost low-fare airlines, competitive airports and an overview of the Latin American situation, in ATRS (ed.), *IX Air Transport Research Society World Conference*, Rio de Janeiro, 3–5 July 2005.

Lipovich, G. (2009). Mercado aerocomercial único en el MERCOSUR; Integración desequilibrada, nuevos procesos y nuevas consecuencias territoriales. *XII Encuentro de Geógrafos de América Latina*, Uruguay.

Lipovich, G. (2010). *Los aeropuertos de Buenos Aires y su relación con el espacio metropolitano*. Buenos Aires, Argentina: Universidad de Buenos Aires.

McKinsey & Company (2010). *Study of the Air Transport Sector in Brazil*. Rio de Janeiro, Brazil: McKinsey & Company.

OAG (2010). *OAG Facts December 2010: Frequency and Capacity Trend Statistics*. Luton, UK: UBM Aviation Worldwide Ltd, Bedfordshire/OAG Aviation. Available at: http://www.oagaviation.com/Solutions/Reports-Guides/OAG-FACTS [accessed: September 2012]

Oliveira, A.V.M. (2005). *An Empirical Model of Low Cost Carrier Entry: The Entry Patterns of Gol Airlines*. São Paulo, Brazil: Center for Studies of Airline Regulation and Competition.

Oliveira, A.V.M. (2006). Patterns of low cost carrier entry: Evidence from Brazil, in Lee, D. (ed.) *Competition Policy and Antitrust*, Vol. 1. Oxford: Elsevier, 297–328.

Oliveira, A.V.M. (2008). An empirical model of low-cost carrier entry. *Transportation Research Part A: Policy and Practice*, 42, 673–95.

Roessing Neto, E. (2007). Perspectivas de um acordo de céus abertos na América do Sul. *Revista Jurídica*, 9, 114–33.

Villena, M.J., Harrison, R. and Villena, M.G. (2008). Impacto Económico de la Política de Acuerdos de Cielos Abiertos en Chile. *Revista de Análisis Económico*, 23, 107–49.

World Airline News (2011). *Gol Announces VARIG will Acquire Webjet*. Available at: http://worldairlinenews.wordpress.com/category/webjet-linhas-aereas/ [accessed: 20 August 2011].

Chapter 6
Low Cost Carriers in Mexico

Alexander O. Scherer Leibold, Yazmín Aguilar López,
Montserrat Flores Zozoaga and Brooke Porter

Introduction

This chapter will discuss the evolution and current status of the airline industry in Mexico, employing case studies of local airlines to support specific points. Air transportation in Mexico has been traditionally regulated by the government. It was not until the year 2005, when the government ceased to regulate in favour of private enterprises, that low cost carriers (LCCs) began operations in the country, allowing new airlines to take part in the air transportation market.

Following the shift in regulations, the commercial airlines operating in Mexico became defined into these groups (also see Figure 6.1):

1. *Regular airlines:* These airlines were the first to operate in the country (e.g. Mexicana de Aviación, established in 1921, and Aeroméxico, established in 1934). After the bankruptcy of Mexicana, Aeroméxico operated the most routes with the greatest connectivity within and outside the country.
2. *Regional airlines:* These were created following the regular airlines; the services focus on domestic flights.

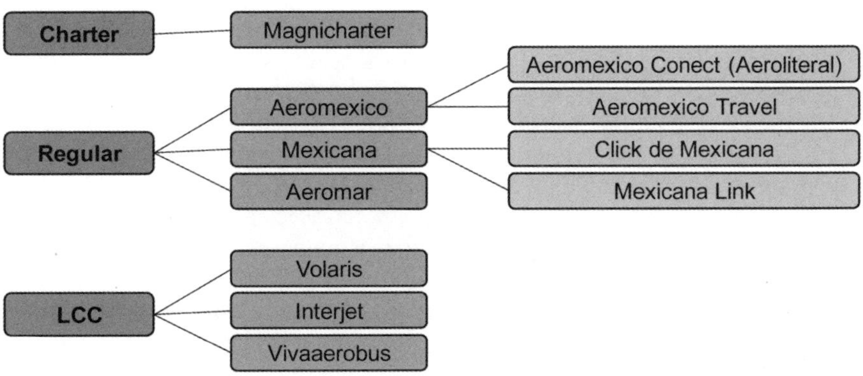

Figure 6.1 Classification of airlines in Mexico

Source: SECTUR 2007.

3. *Low cost carriers:* Operations of LCCs began in 2005 and have since generated an important number of routes and opportunities for travellers by offering lower fares and high quality.[1]
4. *Charter airlines:* The charter market in Mexico is dominated by the tour operator Magnicharter, which sells package tours to various destinations.

The Mexican government has several departments that work to regulate different sectors; among them is the Secretaría de Comunicaciones y Transportes (SCT), whose mission is to promote safe, efficient and competitive transportation and communications systems; contribute to the sustained economic growth and social development of the country; expand coverage and accessibility of services; achieve the integration of Mexicans; and respect the environment. The division that registers and regulates the operations of airlines is the Dirección General de Aeronáutica Civil (DGAC), a part of the SCT.

LCCs first appeared in Mexico when the SCT gave the permission to create new airlines. Some of the first new carriers were Click, Avolar , Volaris, Interjet and Air Madrid. Later, additional companies entered the market, such as Gol (Linhas Aéreas Inteligentes) and Alma (Aerolíneas Mesoamericanas) (Aerolineas Mexicanas, 2008a), affording even more options to the traveller. The first LCC operations began by offering flights to the most visited destinations with lower fares than the regular airlines. Reduced fares were made possible due to the structure of the LCCs, which included the modernization of fleets, allowing for more efficient operations.

LCCs have gained a new market niche. According to SCT, 2,642 million people are using ground transportation in Mexico. LCCs are looking to gain 1 per cent of that market, an equivalent of an additional 2.64 million passengers a year.

Status of Aviation in Mexico

Mexico is the fourteenth largest country in the world, covering nearly 2 million square kilometres. In 2010, its population is estimated to be around 112 million people. It is a representative democratic republic and consists of free states united by a federal pact. Mexico‘s economy is based on an open market aimed at exports. According to Business Monitor, Mexico is the 14th largest economy in the world based on gross domestic product (GDP), (current prices, US$) and the 11th largest according to gross domestic product (at purchasing power parity (GDP at PPP)). In 2010, Mexico's GDP (Current Prices, US$) was US$1.039 trillion with a GDP (PPP) of US$1.567 trillion. By 2011, Mexico‘s economy GDP (PPP) was growing

1 The Mexican Secretaría de Comunicaciones y Transportes (SCT; ministry of communication and transport) include LCCs in the group of regular airlines but those airlines do not distribute their products through a global distribution system (GDS) or travel agencies.

at 5.79 per cent to US$1.658 trillion. Mexico's GDP (PPP) is expected to grow annually by 4.68–5.47 per cent from 2012 to 2016. By the end of 2016, Mexico's GDP (PPP) will reach US$2.115 trillion (Economy Watch 2012).

The airlines are classified according to the services they provide, including those services dispensable to the passenger. Before the advent of LCCs, regular airlines had the entire travellers' market share. LCCs gained a market advantage by offering the passenger existing and new destination options at a lower cost, but in exchange for various services (e.g. meals on board). Regular airlines, seeing the growing competition of LCCs, launched their own subsidiary LCCs in order to compete and regain a part of this market niche.

During 2010, there was registered growth of 3.2 per cent in the number of passengers transported by air, in both national and international operations. Just over 50 million people travelled by air, about 1.6 million more than in 2009. It should be noted that a decrease in passengers travelling by air was observed after 2008; the 2010 figures are about 5.8 million (10.3 per cent) less than the total numbers of passengers in 2008 (CNET-Anáhuac 2011).

One of the main reasons for the decline in tourism in 2009 was the swine flu pandemic (caused by the H1N1 virus): the majority of flights, both domestic and international, showed a drop of the load factor. In 2009 the tourism sector was severely affected (Secretaría de Salud México 2009). Near the end of that year a slow recovery was observed.

Economic phenomena, in addition to security and sanitary issues, also affected activity in the sector. Mexicana Airlines closed operations after reporting a poor financial situation (NOTIMEX 2012). Mexicana previously transported 30 per cent of Mexico's domestic and 20 per cent of international passengers. When the Federal Aviation Administration (FAA) of the United States demoted Mexico to Category 2, this prevented the expansion and change of flight itineraries or routes by Mexican airlines flying to the United States. Category 2 is given to a state or country that fails to meet the guidelines set by the FAA.

The average growth rate between 2009 and 2010 showed little variation (see Figure 6.2). In the case of domestic passengers, the 0.2 per cent growth rate was marginal; international passenger figures showed growth with a 6.3 per cent increase, although the previously predicted growth for this sector was an 8.5 per cent increase. Consistent with these figures, airports reported a moderate increase in the volume of passengers using their facilities in 2010 compared to the previous year. However, according to cost analysis and competition within national aviation by SCT, passenger traffic remained below 2008 levels (Sectur 2011).

The SCT promotes competition in the national market, thus offering the consumer more options and minimizing monopolies in any given region. This has resulted in an overall increase in passenger numbers and, in some cases, a reduction in the fares. The decrease in air fares has opened up air transportation to a segment of the population who previously could not afford it.

Government approval of air fares has been eliminated, giving carriers the freedom to establish competitive prices. However, the results of the new legislation

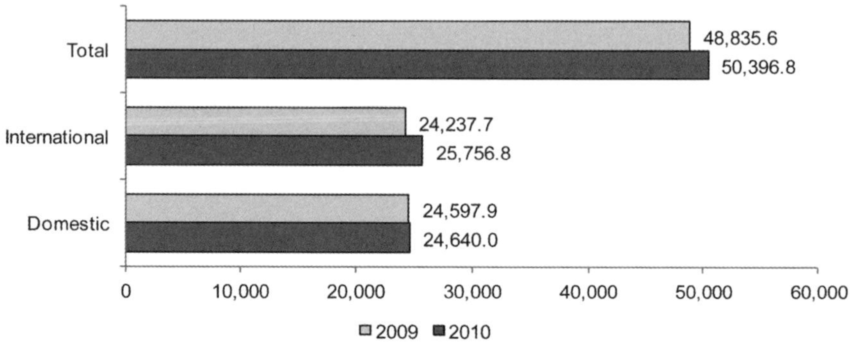

Figure 6.2 International and domestic passenger numbers (thousands)

Note: Based on operating statistics by airline January–December 2009–2010.
Source: Secretaría de Comunicaciones y Transports (2011b).

for air services have not all been positive. Some airlines have failed in recent years (see Table 6.1) due to low levels of liquidity, as the yield was below international standards and the airlines had little access to capital.

Also the cost of leasing aircraft on the Mexican market is inflated, as there is limited or no competition of airport infrastructure in major cities, because there are only four airports managed by groups in Mexico.

Table 6.1 Mexican airlines that have failed since 2007

Airline	Category	2007	2008	2009	2010
Aviacsa	Regular			July	
Aerocalifornia	Regional		July		
Azteca	Regional	March			
Alma	LCC		November		
Avolar	LCC		October		
Aladia	Charter		October		
Nova air	Charter		August		
Mexicana	Regular				August

Source: El Clima 2007.

From 2005 to 2009, changes within the Mexican airline industry included:

1. *Air fleet*: 40.5 per cent increase from 207 to 291 aircraft.
2. *Personnel*: 3.5 per cent decrease from 22,671 to 21,861 people.

3. *Transported passengers*: 27.7 per cent increase from 27 million to 33 million people.
4. *Load factor*: increased 13.20 units, from 55.72 per cent to 68.92 per cent.

It is expected that Mexican airlines will continue to improve the number of passengers on domestic and international flights (CNET-Anáhuac 2011).

Air Travel vs Bus Travel

LCCs use different strategies to gain the confidence of the passengers. Converting bus travellers into air travellers can be challenging. LCCs are able offer seats to a limited number of customers for as little as $500MXP (US$35) to attract consumer attention and confidence. Such low fares easily compete with bus fares and may allow passengers previously limited to bus travel to consider travel by air. These fares are usually made available well before the travel date(s), allowing the airlines to better estimate demand. The difference in fares between LCCs and traditional airlines is made possible due to different load factor strategies. Traditional airlines will fly with a 60 per cent load factor, with higher fares compensating for a potentially lower load factor, whereas LCCs aim for very high load factors by offering tiered fares at competitive prices.

Table 6.2 illustrates the differences between travel by plane and by bus using the Mexico City–Veracruz route as an example.

Table 6.2 Comparison of different modes of transport on the route from Mexico City to Veracruz (July 2011)

Mode of transport	Company	Date of travel	Departure time	Arrival time	Duration of trip	Cost in MXP
LCC	Interjet	3 July 2011	11.45 hrs	12.35 hrs	1 hr	$1,097.25
Traditional airline	Aeroméxico	3 July 2011	11.50 hrs	12.45 hrs	55 min	$2079.00
Bus	ADO platino	3 July 2011	11.00 hrs	16.00 hrs	5 hrs	$622.00

Sources: Information from the websites of airlines and ADO Ground Transportation Company as of 28 June 2011.

The lower fares associated with LCCs may provide additional opportunities for travellers, depending on their needs. For some, lower prices are the most important driver in travel; however, other travellers are limited by schedule. For those needing to travel at a specific time or on a certain date, a LCC may lack the variety of connections that regular airlines or even buses can provide.

Airports

Mexico has 78 major airports, of which 26 offer only domestic flights while 63 provide both domestic and international services. Airport operations are divided between private and public ownership. Of the 78 airports (KAS 2010), nearly half (n=35) are franchised to private entities, 19 are operated by Aeropuertos y Servicios Auxiliares (ASA 2011) and 24 are operated by the government and the state.

The private airport sector is divided amongst Grupo Aeroportuario del Sureste (ASUR 2011), which operates and manages nine airports in southeast México, Grupo Aeroportuario del Pacífico (GAP 2011), which manages 12 airports in the western regions of the country, and Grupo Aeroportuario del Centro–Norte (OMA 2011), which operates 13 airports spread through the country.

Grupo Aeroportuario de la Ciudad de México (GACM) manages Mexico City's International Airport (AICM), one of the largest in the country. Although passenger traffic decreased after 2010 (see Table 6.3), México City's International Airport remains the busiest airport, by passenger traffic, with the airports of Guadalajara and Monterrey following closely behind. It also operates the following airports: Puebla International Airport, Toluca, Cuernavaca Airport and the Airport of Queretaro.

There is also Aeropuertos y Servicios Auxiliares (ASA), a decentralized government organization which manages 19 airports. ASA airports serve many

Table 6.3 Domestic air passenger traffic January–May 2011 vs January–May 2010 (thousands)

Top 10	May 2010	May 2011	var	% var	Jan–May 2010	Jan–May 2011	var	% var
Mexico City	1,366.9	1,516.6	149.7	11.0%	6,253.0	6,567.3	314.3	5.0%
Guadalajara	424.9	404.4	-20.5	-4.8%	1,912.3	1,861.6	-50.6	-2.6%
Monterrey	398.0	417.6	19.6	4.9%	1,769.4	1,827.1	57.7	3.3%
Tijuana	299.1	273.4	-25.8	-8.6%	1,393.2	1,377.9	-15.3	-1.1%
Cancun	291.6	315.2	23.6	8.1%	1,256.6	1,315.1	58.5	4.7%
Toluca	202.4	125.4	-77.0	-38.1%	958.0	747.9	-210.1	-21.9%
Merida	89.5	107.7	19.1	20.2%	424.9	415.3	20.5	4.8%
Culiacan	91.0	98.4	-4.6	-5.0%	436.6	415.3	-21.4	-4.9%
Hermosillo	87.4	89.0	1.6	1.8%	416.6	404.8	-11.7	-2.8%
Tuxtla Gutierrez	53.1	70.7	17.6	33.1%	244.7	312.4	67.7	27.7%
Others	1,054.7	1,034.5	-20.2	-1.9%	4,910.0	4,713.5	-196.6	-0.0
Total	4,358.8	4,440.8	82.0	1.9%	19,975.2	19,988.2	13.0	0.1%

Source: Secretaría de Comunicaciones y Transportes 2011a.

important tourist destinations in the country, including Ciudad del Carmen, Colima, Campeche, Chetumal, Guaymas, and Puerto Escondido. The most important state-managed airports are: Cuernavaca (near Morelos), Celaya (near Guanajuato), and San Felipe (near Baja California). Finally Toluca's International Airport is managed by Administradora Mexiquense del Aeropuerto Internacional de Toluca (AMAIT). This airport is used as a centre for LCCs and as a result has shown a significant growth in operations and passenger traffic (Aeropuertos Mexico 2011).

Overall 76 per cent of total air traffic in the country (both domestic and international) is concentrated in just six airports: Mexico, Toluca, Tijuana, Guadalajara, Monterrey, and Cancun; with 61 per cent of that traffic coming from only three airports (Mexico, Tijuana, Toluca).

Low Cost Carriers

Along with privatization, governments progressed with the deregulation of the markets by eliminating entry barriers and removing price controls. In June 1991, the government deregulated the airline market, mainly through the liberalization of fares and routes, although some restrictions were maintained. Prices were not regulated on routes served by more than one airline; however, the government mandated that prices on monopolized routes would require approval by the regulatory office of the SCT. The SCT also eliminated state-sanctioned exclusivity on routes, and simplified the paperwork required for interested companies to gain access to the market (Sánchez and Somuano 2000).

Following the deregulation of air transport services, the concept of LCCs gained popularity in Mexico, with the following airlines being incorporated into the national market: Volaris, Interjet, Aerobus, Click de Mexicana, Avolar, Alma. These airlines have different operational bases and different numbers of routes (see Table 6.4).

The LCCs are able to pass on competitive fares to passengers as a result of decreased operational costs. LCCs cut budgets by operating with reduced staff and avoiding commissions, that would normally go to brokers and travel agents for ticket printing by utilizing online ticketing systems.

From the year 2005, the concept of LCCs has led to a significant growth of the market. The increase in passengers using this kind of transportation was 38.2 per cent in only two years as a result of lower fares and an increased number of available flights.

Over the past six years, the results of the deregulation of the industry have made sustaining a place in the airline market for many companies increasingly difficult. Only a few LCCs have survived (e.g. Volaris, Interjet, VivaAerobus) while others have declared bankruptcy. One of the problems that remains is the relationship between public and private agencies and between federal and state entities that support the infrastructure and those that allow the development of the country.

Table 6.4 Low cost carriers in Mexico

Name	Volaris	Interjet	VivaAerobus	Click Mexicana	Alma	Avolar
IATA code	Y4	40	VB	QA	CA	V5
ICAO code	VOI	AIJ	VIV	CBE	MSO	VLI
Destinations served in 2007	15	14	24	26	25	17
Direct routes served	56	18	49	22	48	49
Base of operations	TLC TIJ GDL MEX	TLC GDL MTY MEX	MTY	MEX MID	GDL	TJ
Frequent flyer programme	no	no	no	Mexicana GO	no	no

Source: SECTUR 2011.

Market Share

When Mexicana filed bankruptcy, air traffic reduced by 2.5 per cent per month between September of 2010 and January of 2011. Consequently the other companies absorbed a greater part of the market. Specifically, Aeromexico increased its market share from 32 per cent in June 2010 to 45 per cent by November 2010 (Anna.Aereo 2011). Likewise, the main LCC, Interjet, increased its market share from 13 per cent in June 2010 to 23 per cent in June of 2011.

The LCC penetration rate in Mexico's domestic market reached 54 per cent in 2011, a 4 per cent gain from 2010. The total domestic market grew 4 per cent in 2011 to stand at 25.46 million passengers by the end of the year; this figure remains 8 per cent below the peak of 2008 when Mexican carriers transported 27.65 million domestic passengers. The growth observed over the past two years (in 2010 growth was under 1 per cent) is quite an achievement given the sudden collapse in August 2010 of Mexicana, which accounted for a 28 per cent share of the domestic market (CAPA 2011).

Interjet has recorded the fastest domestic growth since Mexicana's exit from the market, now holding a 25 per cent share of the domestic market. According to newly released data from Mexico's DGAC, Grupo Aeromexico remains the largest shareholder of the domestic market at 40 per cent. Interjet was able to gain an impressive 9 per cent of the domestic market share in 2011 by increasing its passenger traffic by 58 per cent to 6.34 million passengers (CAPA 2011).

Meanwhile, LCC Volaris nearly doubled its US traffic during the first seven months of 2011–Mexico market in the period January 2011 to July 2011, compared to a 3 per cent share in the period January 2010 to July 2010 (CAPA 2011).

Interjet has yet to take advantage of the opportunities in the US–Mexico market that have resulted from Mexicana's demise. Although Interjet does not currently operate services to the US, the company has been exploring opportunities to enter the US market and has applied for US Department of Transportation (DOT) authority to serve New York, Miami and San Antonio from Mexico City.

Although VivaAerobus has been serving the US since 2008, more than one year before Volaris entered the market, it did not add additional capacity to the US in the 12 months following Mexicana's shutdown. In fact, VivaAerobus's trans-border traffic decreased in 2011. On its two US routes (Monterrey–Houston and Monterrey–Las Vegas), VivaAerobus transported only 6,600 passengers in July 2011, compared to 8,300 passengers in July 2010. Currently the US market accounts for only 3 per cent of VivaAerobus's total capacity, according to Innovata data.

VivaAerobus has recorded an encouraging improvement in load factor on its US routes, reaching 83.4 per cent in July 2011 after only averaging 67 per cent in 2011. The improvement in load factor is likely a result of the carrier's decision to expand its US network. VivaAerobus planned to add three US gateways beginning in November 2011. VivaAerobus's new services include routes originating in Monterrey to Chicago, Orlando and San Antonio. With these new international routes, VivaAerobus now serves six destinations in the US market, placing it as the Mexican airline with the greatest connectivity with the neighbouring country from Monterrey (VivaAerobus 2012).

Interjet has stated that it plans to continue fast fleet and network expansion despite its decision in late June 2011 to call off a planned initial public offering. The carrier says it is able to continue its expansion using funds from its shareholders and operational profits.

In 2010 most of Interjet's expansion focused on its Mexico City domestic operation. Interjet became the first LCC to serve Mexico City in 2008. It initially operated only eight routes from Mexico City, but the opening of additional slots following Mexicana's collapse allowed it to expand its Mexico City domestic network to 23 cities. Interjet now accounts for 27 per cent of total domestic capacity at Mexico City International, second after the 48 per cent share held by Grupo Aeromexico.

Likewise, Interjet's two LCC competitors, Volaris and VivaAerobus, have also significantly expanded their domestic Mexico City operations since Mexicana's collapse due to the new spaces created in the previously congested airport. Prior to the shutdown of Mexicana, Volaris and VivaAerobus were able to launch small operations at Mexico City, but were limited to only a few daily flights at the coveted, slot-controlled airport.

In 2011 VivaAerobus operated 15 domestic routes from Mexico City with 21 domestic routes at its main hub, Monterrey, and 11 domestic routes at a secondary hub in Guadalajara. Since the shutdown of Mexicana, VivaAerobus has expanded its fleet and now operates 16 B737-300s, compared to its previous fleet of 11 B737-300s. The airline does not have any aircraft on order but by the end of 2011 it had leased four additional B737-300s making a total of 20 B737-300s.

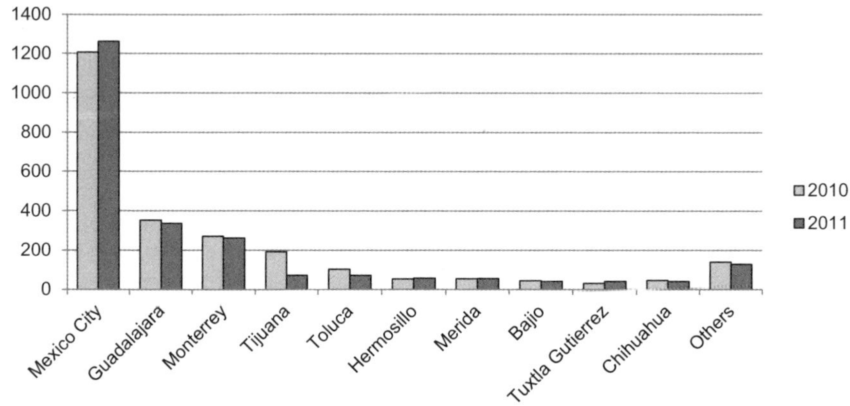

Figure 6.3 Traffic for top ten Mexican airports in (thousand passengers): July 2011 vs January 2010

Source: Secretaría de Comunicaciones y Transportes 2011a.

Volaris currently services 11 domestic routes from Mexico City, its most recent addition being a service from Mexico City to Zacatecas. Volaris runs four routes from its Tolcua headquarters, where it previously had a large operation. Mexicana's demise prompted Volaris to shift several domestic routes to Mexico City. The carrier also has a large operation in Baja California that includes 17 domestic routes from Tijuana (Volaris 2011).

Volaris now operates 31 A320 family aircraft, up from 21 aircraft a year ago. Volaris, like Interjet, Aeromexico and to a much lesser extent VivaAerobus, decided to accelerate fleet expansion in the aftermath of Mexicana's exit. More rapid expansion is expected as Volaris currently has another 17 A320s on order (Volaris 2011).

Overall, Volaris and VivaAerobus have not expanded in the domestic market as rapidly as Interjet. Instead, Volaris has directed a large portion of its additional capacity to the US market, only recording a modest 24 per cent increase in domestic traffic for July 2011 to 423,000 passengers. VivaAerobus saw its traffic grow by 30 per cent in July 2011 to 336,000 passengers. Both carriers have seen small increases of three percentage points in their domestic market shares over the last year, from 14 per cent to 17 per cent at Volaris and from 10 per cent to 13 per cent at the smaller VivaAerobus (CAPA 2011). Figure 6.3 shows the traffic for the top ten Mexican airports.

Case Study

By 2011 VivaAerobus had become the leader in Mexico's LCC market, operating 47 domestic routes. Of these domestic routes 10 are among the 30 busiest routes in the country and 16 are exclusive to VivaAerobus, many of which are new routes for the country. VivaAerobus also operates two international routes.

By 2012 VivaAerobus was flying 58 domestic routes and 6 international routes between Mexico and the US (San Antonio, Houston, Las Vegas, Orlando, Miami y Chicago), and in its first four years of operations, VivaAerobus transported more than 7 million passengers. Its operative strategy generates important savings that are passed on to the passengers, allowing it to offer the lowest fares in the Mexican aviation industry; additionally VivaAerobus offers services such as hotel packages through its web page. VivaAerobus plans to quickly expand its fleet by adding an additional 30 airplanes by 2013 (VivaAerobus 2013).

LCCs are an option for travellers, depending on their needs: for some, the lower prices are the most important characteristic, but for those people that need to travel at a given time, LCCs might not have the variety or frequency of connections that regular airlines or even buses can provide.

Conclusion

During the last decade the aviation industry in Mexico has been subject to several changes, especially in the way it has been marketed (CNNexpansion 2008). The domestic airline deregulation has allowed the entry of new schemes and companies that have brought quality services at lower prices to passengers. However, LCCs in Mexico remain in their infancy compared to LCCs in North America and Europe.

To survive in the Mexican market, LCCs have reached their position by adhering to the rules of the regular airlines. In 2005 it was thought that deregulation would offer better options to passengers by bringing domestic air alternatives. However, the strongest airlines in the country, Aeromexico, instead maintained a perfect oligopoly by distributing the routes among themselves.

The stronger airlines forced some of the smaller airlines into bankruptcy and others to look for new elements to offer the consumer a better service. VivaAerobus has the most similar operation to a classic LCC. It has continued to offer short point-to-point services, thus gaining ground in Mexican transportation.

Predictions for the future are difficult. Though Mexican airlines have complied with international standards, Mexico still must work to develop its domestic transportation industry and continue to respond to government deregulations. This development and growth will not be possible for a single company or group; it must be a synergy between all actors to be successful.

Only one year after the suspension of services at the largest airline casualty since the onset of the global financial crisis, Grupo Mexicana, the Mexican market appears to have fully recovered. Mexico's remaining carriers have been able to quickly absorb Mexicana's previous share of approximately 27 per cent of the domestic market. To date, US carriers have benefited the most from Mexicana's demise. However, Mexico's remaining carriers are eager to start narrowing the very wide gap with their foreign competitors in the international market (CAPA 2011).

Bibliography and References

ADO (2011). Grupo ADO Available at http://www.ado.com.mx/ado/index.jsp [accessed: 28 June 2011].

Aerolíneas Mexicanas (2008a). Alma de México. Available at: http://www.aerolineasmexicanas.com.mx/content/view/63/28/ [accessed: 6 February 2012].

Aerolíneas Mexicanas (2008b). Avolar. Available at: http://www.aerolineasmexicanas.com.mx/content/view/64/28/ [accessed: 6 February 2012].

Aeropuertos Mexico (2011). Toluca International Airport. Available at: http://www.aeropuertosmexico.com/Ingles/content/view/280/380/ [accessed: 15 May 2010].

Anna.Aero (2011). Market Share, 20 April 2011. Available at: http://www.anna.aero/2011/04/20/aeromexico-now-biggest-mexican-domestic-airline/ [accessed: 6 February 2012].

ASA (2011). Aeropuertos y Servicios Auxiliares. Available at: http://www.asa.gob.mx/wb/webasa/weba_aeropuertos [accessed: 15 May 2011].

ASUR (2011). ASUR (Aeropuertos del Sureste) website start page. Available at: http://www.asur.com.mx/asur/index.asp [accessed: 20 May 2011].

CAPA (2011). A year after Mexicana's exit, Aeromexico, Mexican LCCs and US carriers are main beneficiaries. Centre for Asia Pacific Aviation. Available at: http://www.centreforaviation.com/analysis/a-year-after-mexicanas-exit-aeromexico-mexican-lccs-and-us-carriers-are-main-beneficiaries-58181 [accessed: 6 February 2012].

CNET-Anáhuac (2011). Panorama de la actividad turística en México, México: Consejo Nacional Empresarial Turístico (CNET).

CNN Expansión (2008). *Aeromexico va contra las de bajo costo*. Available at: http://www.cnnexpansion.com/negocios/2008/05/21/aeromexico-contra-las-de-bajo-costo [accessed: 15 November 2011].

Economy Watch (2012). Mexico Economic Forecast: Business monitor online. Available at: http://www.economywatch.com/world_economy/mexico/economic-forecast.html [accessed: 14 June 2012].

El Clima (2007). Aerolíneas de bajo costo. Available at: http://www.elclima.com.mx/las_aerolineas_de_bajo_costo.htm [accessed: 15 November 2011].

GAP (2011). Grupo Aeroportuario del Pacifico website. Available at: http://www.aeropuertosgap.com.mx/english/index-site.html [accessed: 15 May 2011].

Grupo Aeroportuario del Centro Norte, OMA (2011). Available at: http://www.oma.aero/es/aeropuertos/ [accessed: 15 May 2011].

Interjet (2011). Interjet website. Available at: http://www.interjet.com.mx [accessed: 4 June 2011].

KAS (2010). *Informe Foro Infraestructura del transporte*. Konrad-Adenauer-Stiftung. Available at: http://www.kas.de/mexiko/es/publications/search/?901_1_1567=INFRAESTRUCTURA+AEREA&form_id=901&901_1_1568_day=15&901_1_1568_month=6&901_1_1568_year=2010&901_1_1569_

day=15&901_1_1569_month=6&901_1_1569_year=2011 [accessed: 15 June 2011].

NOTIMEX (2012). *Solicitan detalles de pérdidas económicas por cierre de Mexicana*, 2 February 2012. México: NTMX: Agencia de Noticias del Estado Mexicano. Available at: http://sdpnoticias.com/nota/293201/Solicitan_detalles_de_perdidas_economicas_por_cierre_de_Mexicana [accessed: 6 February 2012].

Sánchez, F. and Somuano, A. (2000). Privatization, Deregulation and Competition: Evidence from the Mexican Airlines Industry. *Boletín Latinoamericano de Competencia*, 78–86.

Secretaría de Comunicaciones y Transportes (2011a). Dirección Aeronáutica Civil, 1 December 2011. Available at: http://www.sct.gob.mx/transporte-y-medicina-preventiva/aeronautica-civil/ [accessed: 10 December 2011].

Secretaría de Comunicaciones y Transportes (2011b). Estadisticas de Operación por aerolínea. Available at: http://www.sct.gob.mx/uploads/media/AERO PUERTOS_MAYO_11.pdf [accessed: 10 December 2011].

Secretaría de Comunicaciones y Transportes (n.d.). Boletin Mayo 2011. Available at: http://www.sct.gob.mx/uploads/media/Boletin_May_2011_es.pdf [accessed: 10 December 2011].

Secretaría de Salud México (2009). México Sano. Available at: http://portal.salud.gob.mx/descargas/pdf/period_mexsano/mexicosano_jun09.pdf [accessed: 6 February 2012].

SECTUR (2007). *Líneas aéreas de bajo costo amplian su mercado*. México: Secretaría de Turismo. Available at: http://www.aerolineasmexicanas.com.mx/content/view/85/27/ [accessed: 10 December 2011].

SECTUR (2011). *Análisis Estratégico de las Líneas Aéreas de Bajo Costo en México. (ABC)*. Available at: http://www.sectur.gob.mx/wb/sectur/sect_Analisis_Estrategico_de_las_Lineas_Aereas_de_ [accessed: 10 December 2011].

VivaAerobus (2011). VivaAerobus. Available at: http://www.vivaaerobus.com [accessed: 4 June 2011].

VivaAerobus (2012). VivaAerobus website. Available at http://www.vivaaerobus.com/mx/noticias [acessed: 12 June 2012].

Volaris (2011). Volaris website. Available at: http://www.volaris.com [accessed: 4 June 2011].

PART IV
Asia and Oceania

Chapter 7
Low Cost Carriers in Asia and the Pacific

Semisi Taumoepeau

Politics, Societies and Cultures in Asia

At the start of the twenty-first century it was confidently predicted that this period would become the 'Asian Century'. The political landscape in Asia is constantly evolving. Political dynamics within Asian states range from vibrant multiparty democracies to some of the world's most closed and repressive regimes. Historical mistrust among states in the region has the potential to explode long-dormant rivalries into conflict, while at the same time the proliferation of cooperative multilateral mechanisms offers new opportunities for peace (National Bureau of Asian Research 2011).

Understanding the personalities, historical influences and contemporary issues that inform the political landscape in Asia is of critical importance in understanding the culture and social factors that affect economic growth in this region. The societies and cultures of the Asia–Pacific region are undergoing enormous changes which have radically altered the life of the average person, making it very different from that of their parents' and grandparents' generations. In most societies in the region, there are tensions between tradition and modernity, especially how Western and other Asian cultures have had an impact on traditional cultures. The growing economy in this region plays a significant role in how societies in the region are changing. Issues which affect people's everyday lives, include the family, gender relations, religion, the arts, film, ethnic relations, population migration education, and images of the Asia Pacific region (Maidment and Mackerras 1998).

Economic Growth in the Asia Region Continues

Both the World Bank and the Asian Development Bank are predicting continual economic growth for most Asian countries for 2013–2014. Growth in developing East Asia for the same period will remain slow, mainly due to weakening external demand. Global growth has also been affected by supply shocks from geopolitical disturbances in the Middle East, supply chain disruptions following the earthquake and tsunami in Japan, and a slower-than-expected recovery of private demand in crisis-affected countries.

The economic growth for the Asia region continues. The proportion of people living on less than US$2 a day in developing East Asia is expected to remain at about 24 per cent for 2013, as in the previous two years. Poverty reduction efforts would be hampered in the event of another sudden increase in food prices against a backdrop of slowing income growth. Floods in Thailand and the tsunami in Japan have affected the entire region as the impacts of the disasters have spread through the industrial supply chains. While reconstruction after the flood in 2012 contributed to growth, the resilience of East Asia's production networks is being tested once more (World Bank 2013).

Led by China, aggregate sub-regional gross domestic product (GDP) growth will moderate somewhat from its 9.6 per cent recovery pace in 2010 to about 8 per cent in 2013. Slower investment and less export growth will slow down expansion in China – from the 10.3 per cent recovery to 9.6 per cent in 2011 and 9.2 per cent in 2012. Hong Kong, China; Taipei, China, and the Republic of Korea will settle back to more sustainable growth of around 4 per cent in 2013.

In Southeast Asia, GDP will expand by 5 per cent in 2013. This moderation is most notable in Malaysia, the Philippines, Singapore, and Thailand, which grew rapidly during the recovery. The region is also playing its part in rebalancing growth toward private domestic demand, as seen in the increased investment rate in Indonesia. In South Asia, after growth or 7.9 per cent in 2010 in line with the global recovery, growth will continue steadily in 2013. Leading the sub-region, India's 2010 expansion was robust and broad-based (8.6 per cent) – and is set to remain strong at 8.0 per cent in 2012–2013 (Asian Development Bank 2013).

Resource-rich Economies Will Drive Growth in the Pacific

A 6 per cent growth forecast for 2013 is attributed to the resource-rich economies of Papua New Guinea, Timor-Leste and Solomon Islands, which are benefiting from higher global commodity prices, new investment, and higher government revenue from resources. While income from tourism and remittances generally picked up in step with the global recovery, most small Pacific economies are projected to grow only slowly (2 per cent or less for 2013–2014). Inflation is also picking up alongside global commodity prices as these import-dependent countries cope with rising world food and oil prices. Inflation is forecast to rise above 6 per cent in 2013 (Asian Development Bank 2013).

Evolution of Low Cost Carriers in the Asia Pacific Region

This chapter examines the development and structures of selected low cost carriers (LCCs) in the Asia Pacific region. Traditionally aviation in the Asia Pacific region has been regulated. Even up to the early 2000s, the aviation industries of most countries in the Asia Pacific region were still highly regulated, dominated by a few long-established national airlines (owned or partly owned by their respective governments) such as Air India, Garuda Indonesia, Singapore Airlines, Malaysia Airlines and Thai Airways (O'Connell and Williams 2006).

However, the economic crisis of the early 1990s partly forced some of the Asian countries to liberalize their aviation policies. Deregulation started a revolution in the global aviation market. It effectively opened up new air services and removed government control (Doganis 2006, Taumoepeau 2010a and ICAO 2009). Countries in South East Asia were quick to recognize the potential commercial benefits from some form of deregulation and adoption of the LCC model, a genuine product of the market deregulation process that started from their North American and European counterparts (Papatheodorou 2010). Asian national airlines and private companies subsequently – in the early 2000s – either formed LCC subsidiaries or established LCC companies. India and Japan followed suit with their own brand of LCC models few years later. China also recognized the need for a more liberal aviation regime and the subsequent commercial benefits it brings, but focused first on its centralized aviation policy and primed up the operation of the three main national airlines, albeit under the Civil Aviation Administration of China (CAAC), which is the aviation authority under the Ministry of Transport of the People's Republic of China. As a result, their first LCC model Spring Airline is still evolving as a hybrid model.

The LCCs in the Asia Pacific region, which have successfully brought down airfares in most countries, has managed to empower millions of its inhabitants to travel. This, coupled with a growing middle class in the region, has caused an unprecedented level of air travel domestically, regionally and internationally.

The LCC seems to be built for economic recovery and recession, has managed to break through bottlenecks and open new business opportunities, and has unlocked potential for economic growth regionally. The Asia–Pacific Economic Cooperation (APEC) member countries are watching with interest the development of LCCs in the Asia Pacific region. APEC members account for about 40 per cent of the world's population and about 50 per cent of the world's GDP and trade. They actively facilitate more trade and tourism activities overall and empower the main bulk of the population to afford air travel both domestically and internationally in Asia and the Pacific.

Asia Pacific is a diverse geographic region, it is home to 4 billion people, i.e. about 62 per cent of the world's population, and its peoples have a very wide range of income levels. The region generates 30 per cent of global GDP, which boosts economic growth globally. Aviation is widely recognized as a key contributor to economic and social development in the region.

Most countries support the liberalization of air services, which allows for more competition in the market to the benefit of consumers. During the past few years there has been much development in liberalization: bilateral aviation agreements have been established; new aviation regulations have been adopted in the bigger economies such as India and China; and the proliferation of new carriers, especially low cost airlines (LCAs) in the region and beyond.

The impact of oil prices on growth in the Asia Pacific region is still very significant due to the intensity of energy production in the region. Much of the airport development in the Asia Pacific region is still taking place in China and India, though there are notable projects in Vietnam, the Philippines and Indonesia, for which private sector participation is being sought in these three countries.

The Asia Pacific carriers' share of the global market is 28 per cent of global passenger traffic and 43 per cent of global cargo traffic. The Asia Pacific carriers generate more than US$156 billion of revenue, carrying 619 million passengers, of which 433 million are domestic and 186 million international, and transporting 18 million tonnes of cargo.

The Pacific Islands market is characterized by small economies, limited natural resources for exports, and widely dispersed populations spread across many isolated islands. The provision of air services is fragmented, often involving long routes with thin traffic and low freight levels. As a result, Pacific airlines face considerable constraints in managing returns through low passenger and cargo levels and achieving sufficiently high levels of aircraft utilization. The more commercial Pacific air routes, with closer proximity to Australia and New Zealand, are generally tourism-related (see Chapter 9). Tourism has been an important contributor to the economic growth of these islands (Asian Development Bank 2007).

The introduction, beginning in 2004, of LCAs such as Pacific Blue, Polynesian Blue and Freedom Air into seven Pacific routes from Australia and New Zealand produced dramatic passenger growth and substantial economic benefits. It also resulted in considerable competitive pressure on fare levels to the benefit of consumers. The Fiji Islands, Samoa, Tonga and Vanuatu were the markets targeted by LCAs. Passenger traffic and visitor arrivals in all four target countries increased substantially (Asian Development Bank 2007, Taumoepeau 2010a). Coinciding with this increase in competition on some routes, air service provision and stability in the Pacific generally improved, despite continuing high aircraft fuel prices in the region.

Development and Structures of Low Cost Carriers in the Asia Pacific Region

Bilateral to Multilateral Air Services

Since the end of World War II, international air services were operated between countries under strict bilateral air service agreements (ASAs) negotiated between the two countries. Typically, these ASAs specified which airlines can fly between two countries, the routes, the airports, whether airlines could fly beyond services (fifth freedom rights), frequency and capacity operated. In addition to the ASAs, the countries would have restrictions on ownership and control, limitation on foreign ownership and some defence, safety and strategic issues. However, in the

Figure 7.1 The emergence of Tiger Airways was through a subsidiary formation of an existing national airline, Singapore Airlines
Source: Michael Lück.

last two decades there has been a global trend to liberalize some of these aviation requirements as countries recognized the benefits of allowing market forces to take over flights, trade, tourism and international investment. Liberalization of air services started in the USA with the Airline Deregulation Act of 1978, which pursued an active policy of liberalization through the so-called 'open skies' agreement at an international level, followed by the European Union (EU) in the 1980s, and it is now being adopted in the Asia Pacific region (Graham, Papatheodorou and Forsyth 2010).

As North America learned from the Southwest Airline model, recent movements in the Asia Pacific region for deregulation of once highly regulated aviation markets have initiated the emergence of LCCs in this region. The first LCC was Malaysia's Air Asia, established in 2001, and it was quickly followed by other carriers. These carriers were established either through a subsidiary formation of an existing national airline such as Tiger Airways (Singapore Airlines – Figure 7.1), or as a newly formed LCC such as Spring Airline in China, formed in 2005, or Air Deccan, founded in India at the same time.

Table 7.1, on the following page, shows stages and milestones in the evolution of aviation policies from highly regulated regimes to deregulated airline practices from the late 1970s to the early 2000s. The Asia Pacific region adopted the LCC evolution in the late 1990s, much later than the USA and Europe, and it continued

Table 7.1 Evolution of airline practice key factors from the late 1940s to the early 2000s

Region	Late 1940s–1970s	1980s	1990s	Early 2000s
USA	Bilateralism	Multilateralism LCCs ICTs Alliances	Multilateralism LCCs ICTs Alliances	Multilateralism LCCs ICTs Alliances
Europe	Bilateralism	Bilateralism ICTs	Multilateralism LCCs ICTs Alliances	Multilateralism LCCs ICTs Alliances
Australia and NZ	Bilateralism	Multilateralism Alliances ICTs	Multilateralism LCCs ICTs Alliances	Multilateralism LCCs ICTs Alliances
East Asia and Southeast Asia	Bilateralism	Bilateralism	Bilateralism Alliances ICTs	Multilateralism LCCs ICTs
South Pacific	Bilateralism	Bilateralism	Bilaterialism Alliances (Codeshares)	Bilateralism ICTs Alliances (Codeshares)

Notes: LCCs: Low Cost Carriers; ICTs: Information and Communication Technologies.
Source: Taumoepeau 2009.

into the mid-2000s. Multilateralism took over from bilateralism as the deregulation process progressed through the Asia Pacific region.[1]

An outstanding feature of the emergence of LCCs in the Asia Pacific region is how they managed to evolve into a variety of LCC models (some are very different from the original Southwest Airline model of the 1970s) to suit their respective geographical location (see Table 7.2), market needs, restrictive aviation regimes and varying stages of tourism and infrastructure development. LCC operators in the region compete with the traditional national carriers, embrace tourism growth in these regions and address cumbersome challenges of flying commercially above the biggest stretches of ocean in the world: the Pacific. Finally, despite the many different stages of economic development and many geographical borders, the LCC has been successful in lifting the level of domestic travel and international tourism receipts in the region.

---

1 A bilateral air transport agreement is a contract between two countries or states. This allows the airlines of both countries or states to operate commercial flights that cover the transport of passengers and cargoes of both. A multilateral air services agreement is the same as a bilateral agreement; the only difference is that it involves more than two contracting countries or states.

The LCC in Asia typically evolved either as a subsidiary of an existing traditional national airline, such as Tiger Airways from Singapore Airlines, or Air Asia from Malaysia Airlines, or as a newly established company such as Air Deccan in India. Neither model has any particular advantage over the others,

Table 7.2 Low cost carriers: Mature markets vs emerging markets

Mature markets	Emerging markets
(USA, Europe) sluggish economic recoveries and cautious consumers	(Asia Pacific) emerging economies and their citizens' soaring income
Tend to be saturated, look for new segments or fly further (very common) e.g. easyJet from London to Amman, Southwest and jetBlue flying transcontinental, Ryanair to Mediterranean	Far from saturation, fly further (not common yet) for the international market. Trying to expand beyond home markets and around the region. Indonesia's Lion Air and Philippines' Cebu Pacific restrict themselves to domestic market. AirAsia, Jetstar Asia and Tiger Airways expanding to other countries in the region
Benefit of large common aviation area (market)	*No benefit from large common aviation area (market)*
e.g. Europe	e.g. Asia region. Air Asia develops in the Asia region by creating different structures to suit each country's aviation requirements. China does not allow LCCs commercial freedoms such as negotiation with airports
Financial results after 2009 recession	*Financial results after 2009 recession*
Maintain double-digit growth	Maintain double-digit growth
Strategy for expansion	*Strategy for expansion*
Demographically and geographically	Geography, first time customers (take some road transport customers)
Take up charter carrier markets (e.g. Spain)	
Flying further (change the business model)	Domestic and also flying internationally
Looking for more and new segments	Looking for more segments as Phase 2
Distribution via GDS	Start distributing via GDS
More revenue from ancillary sales	More revenue from ancillary sales
Becoming a hybrid LCC: add products (with some cost). Traditional carriers cut out frills and costs. Converging in all directions.	Still on single fleets Dual brand strategy e.g. Singapore Airlines and Tiger Airways
Cooperation with network carriers. Air Berlin joining oneworld Tend to go into dual fleet e.g. Virgin Australia	A LCC may have different structures in different markets it serves, depending on the aviation environment, competitions, market characteristics, etc. (e.g. Thai Airways and Thai Tiger, Nok Air and Thai Wings)
Towards end of golden age of lower fares	*Prospects for golden age still high*
Air passenger duty and emissions trading	

Sources: Airline Business 2011, CAPA 2011, and prepared by the author.

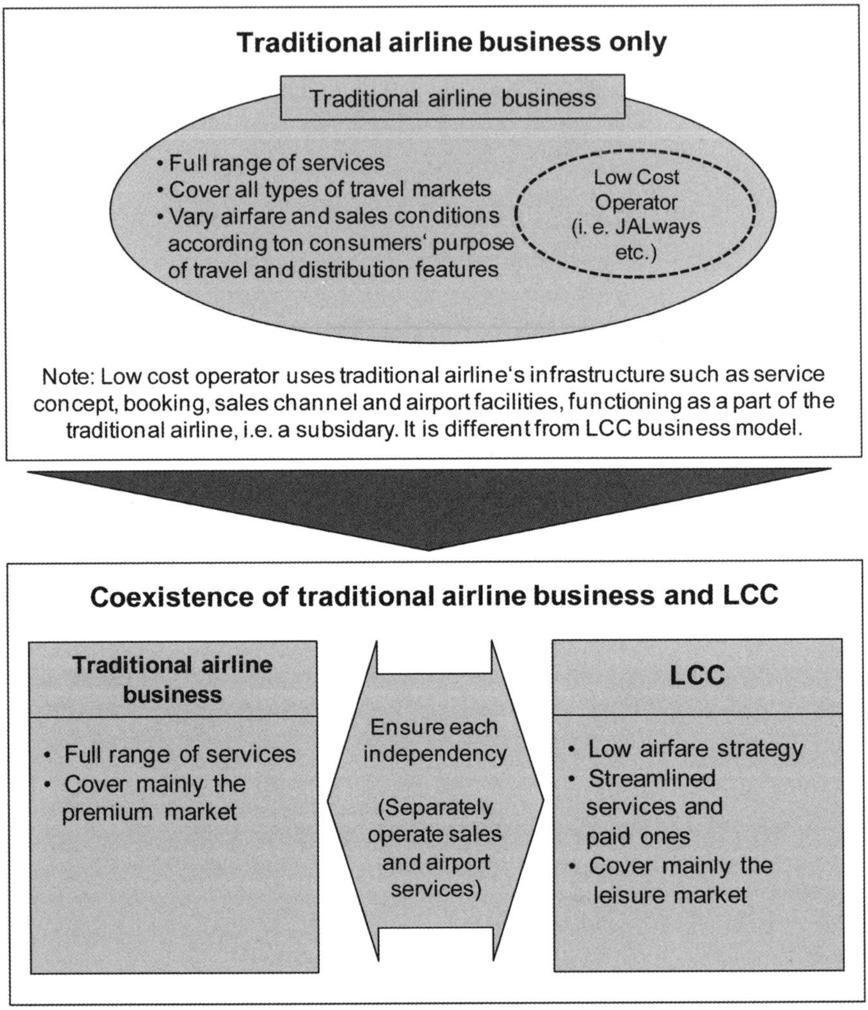

Figure 7.2 Traditional airline business model vs new low cost carrier model
Source: Wijaya 2010.

all depend on the airline's mission, area of operation and home market, and the existing aviation policies and regulations governing the region (see Table 7.2).

Figure 7.2 shows a distinction between a low cost operator arising from an existing traditional airline business and a start-up LCC model in coexistence with a traditional airline. In addition to these two typical structures, in the Asia Pacific region there are several variations of LCC models, as characterized by the following adopted models depending on special circumstances, unique to air routes and markets served:

1. LCC hybrid model and still subject to highly regulated aviation regime (e.g. Spring Airline in China).
2. LCC universal model in a recently partly deregulated aviation regime (e.g. IndiGo).
3. LCC universal model in a fully fledged liberalized aviation regime (e.g. Tiger Airways in Singapore and Air Asia in Malaysia).
4. LCC Pacific model operating from the metropolitan centres (e.g. Jetstar, Pacific Blue in Australia and New Zealand).

Institutional Framework

Governments in the Asia Pacific region, which once guarded airspace under the rigid principles of traffic rights reciprocity, now have to show 'flexibility' as LCCs change the rules of the game. Open skies agreements in Asia are now becoming common (IATA 2010). LCCs have tough booking conditions and cancellation terms, which are often in conflict with regular passenger rights. Directly or indirectly, national carriers will do everything possible to ensure that LCCs do not take away significant market share. In addition, aviation authorities, after years of having 'protected' the interests of the national carriers and safeguarded them from competition, will need to ensure that national carriers do not use their power to deliberately displace LCCs (IATA 2010, Turner and Witt 2011, CAPA 2004).

After the death of Chairman Mao in China in 1976, his successor, Deng Xiaoping, advocated reforms aimed at ending China's economic and political isolation. He realized the importance of tourism and air travel and the need to open up China to attract foreign exchange for development. The aviation industry of India, was similarly heavily regulated up to the early 1990s with a government that owned Air India and dominated both the international and the domestic aviation scene. The production of a roadmap for the reformation of India's outdated aviation policy in 2003 opened up the airways for LCCs to take off in the country.

However, the aviation situation in the Pacific is different. With the exception of the Territories of France and New Zealand, air transportation is still regulated in most Pacific countries, which still cling to their traditional bilateral air service rights, despite a recent move initiated by the Pacific Forum (2004) to open up to several ASAs. The small national carriers of the Pacific still rely on traditional rights to gain access to neighbouring destinations, as well as to Australia and New Zealand, which constitutes their bread-and-butter business, supplying them with much-needed tourist traffic and foreign currency. Due to their size and very small economies, most of these islands and their national carriers are unable to switch to a LCC-type operation. The exception is the Samoan government, which set up a new LCC model with Pacific Blue, owned by Virgin Australia, in 2005. Thus the main LCC in the South Pacific is Virgin Australia. Traditional carriers such as Air Pacific, Air Vanuatu, Air Niugini, PNG airlines, Solomon Airlines and Our Airline are still operating under the old national government-owned model (see Figure 7.3). Others, such as Royal Tongan Airlines and Polynesian

Figure 7.3 Traditional carriers such as Our Airline are still operating under the old national government owned model

Source: Michael Lück.

Airlines ceased operation in 2005, mainly because of their inability to adopt an economically sustainable model. Their geographical isolation, small economies, limited tourism infrastructure and small accommodation capacity mean that the concept of having a home-grown LCC model (which would need a steady flow of traffic), or a variation of it, may not work in the Pacific Islands at this point in time (Taumoepeau 2010b).

Highest Aviation Growth Region in the World

In 2009, passenger traffic within Asia Pacific went up to 647 million passengers, making it the highest growth region in the world in terms of international tourism (PATA 2011), compared with 638 million in North America. In Southeast Asia alone, more than 20 LCCs were established, compared with just a handful of carriers formed just a decade earlier. Asia's LCCs have greatly increased their market share over the traditional airlines during the recent economic crisis (PATA 2011).

The number and proportion of LCC aircraft is likely to grow even more, as Indian – and eventually Japanese, Korean and Chinese – LCAs enter new markets, and as liberalization spreads across the fast-expanding Asian market. Indonesia's Lion Air (a low cost full service airline (FSA)), which holds a 40 per cent domestic

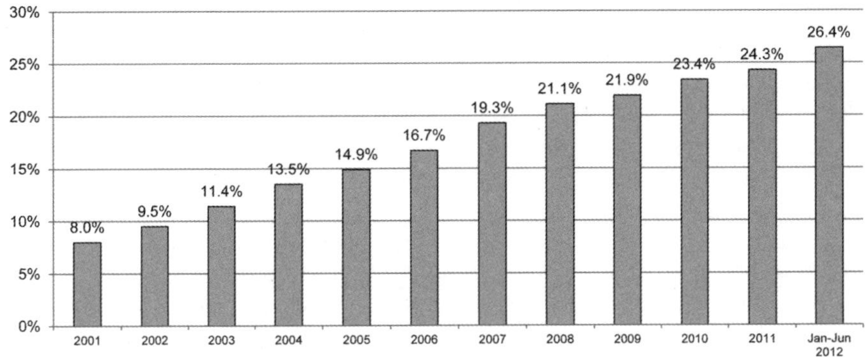

Figure 7.4 Low cost carrier capacity share (%) within the Asia Pacific region 2001–2012

Source: CAPA 2012, with data provided by OAG, a UBM Aviation business.

share) is planning to spread its wings through Asia, with two A320s due to arrive each month for the next four years; and AirAsia ordered 200 Airbus 320s in 2011.

The market share (in terms of seats) of LCCs within Asia reached 19.1 per cent at the end of 2011 (see Figure 7.4) and is expected to reach 20 per cent by the end of 2011 (CAPA 2011). Those market gains have come at the direct expense of Asia's traditional airlines. LCCs tend to respond more quickly to economic recessions and customer preferences by changing schedules or operational cost structures much more swiftly than the established traditional airlines.

However, LCC penetration in the region is still behind that of the USA (28 per cent) and Europe (37 per cent), while the middle class across Asia is growing rapidly, creating new demand for affordable air travel. Furthermore, the upcoming open skies policy among the Association of Southeast Asian (ASEAN) countries, as well as increasing liberalization in Japan and South Korea, will further boost air travel in the region.

These new market opportunities, plus strong competition from AirAsia, Cebu Pacific, Lion Air, IndiGo, Spring, Skymark and others, has led several FSAs in the region to set up their own low cost subsidiaries in recent years (for example Qantas with Jetstar, Singapore Airlines with Tiger, Malaysian Airlines with Firefly and Korean Air with JinAir). Joining these airlines in the next year will be Thai Airways and All Nippon Airways, while Singapore Airlines just announced plans to establish a long haul LCC.

Even in a year as tough as 2009, double-digit growth has remained the norm in the fast-growing Asia-Pacific low cost sector. This was aided by a developing market and lower exposure to the economic crisis. Some of the strongest progress has been in the short haul market around Southeast Asia, India and Australia. AirAsia and its Thai and Indonesian associates, Tiger Airways and Jetstar Asia, are some of the successful LCCs growing their market share in Southeast Asia.

Some double-digit growth has also been seen in India, Japan and, recently, in the relatively small but evolving low cost sector in China.

The low cost model in the Pacific is somewhat different from its Asian counterpart. The region has smaller economies, fewer inhabitants and, with islands sparsely isolated throughout the vast Pacific Ocean, poses a different set of challenges to adjust to, not only for the low cost model but also the traditional national network airlines.

In the Pacific region, aviation links to the South Pacific islands were provided by national airlines from Australia, New Zealand and the Pacific Rim countries (Kissling 1998). Shortly after, most South Pacific countries gained their political independence and small national airlines evolved (Forsyth and King 1996, Kissling 2002). The aviation industry in the whole Pacific region (except the American state of Hawaii and the USA territories in the North Pacific) is still highly regulated. Traditionally, air services from the metropolitan gateways, Sydney and Auckland, have dominated air transportation in the region since colonial days. Independence status accorded to several Pacific countries in the 1970s paved the way for some bilateral ASAs to be put in place and also facilitated the establishment of several small national airlines in the Pacific. From the early 1980s air transportation and tourism became very much an integral part of the growing economies of many countries in the Pacific. Aviation regulation in the South Pacific showed some bias towards bilateral regimes to protect and enable individual growth for the small airlines. Support from host governments is still needed, as national airlines continue to play a public utility service role. The level and degree of government involvement in the national airlines and tourism development varies depending on government objectives and stages of economic development. The small national airlines of the Pacific are still actively exploring codesharing, alliance opportunities and even some aspects of LCC models. Airports and runways in the region are not uniform in standard and runway lengths, and some are inadequate for A320 and B737 operations (Taumoepeau 2007).

In terms of aircraft for the LCA sector in the Asia–Pacific region, hundreds of new aircraft are being ordered by LCCs, many more than by the national airlines, signalling a healthy growth period for the LCC sector by comparison with the more traditional airlines in the region. India's GoAir (72 new Airbus A320s), IndiGo (180 new A320s) and Cebu Pacific of the Philippines (37 new A320s) had each signed deals with Airbus. It is confidently expected that the LCCs could easily capture up to half of all air travel in the Asia Pacific region within the next 20 years. More than 40 LCCs were established in the region from the late 1990s to the early 2000s, with more than 847 aircraft, more than 80 per cent of which are B737s or A320s. Jetstar operates 46 Airbus A320s and Virgin Australia operates 10 Boeing B737s within Australia and the Pacific Islands (see Table 7.3).

Malaysia's AirAsia, which sparked the growth of budget travel in the region, confirmed in 2012 a new deal with Airbus for up to 200 aircraft. AirAsia chief Tony Fernandes said that this boosted the firm's Airbus fleet fivefold to 500 aircraft as more people in the region demand cheaper flights. 'We have 600 million people

Table 7.3 Low cost carriers based in the Asia Pacific region 2010

Country	Low cost carrier	Year commenced operations	Fleet size (e.g. A320/330 or B737)	Destinations/region
China	Spring Airlines	2004	28 A320	China, Hong Kong, Macau
	China United	1986	14 B737-800	China Domestic
	Hong Kong Express	2004	5 B737-800	China, Hong Kong, Japan, South Korea
India	Air-India Express	2005	21 B737-800	India Dubai Himalayas
	GoAir	2005	12 A320	India Domestic
	IndiGo	2006	51 A320	India Dubai Bangkok Singapore
	Indus Air	2006	2 CRT200ER	India Domestic
	jetLite	1993	19 B737-800	India Nepal
	jetKonnect	2003	8 ART72-500, B737-800	India Domestic
	Kingfisher Red	2003	21 B737	India Sri Lanka Dubai
	SpiceJet	2005	40 B737	India Kathmandu
Indonesia	BataviaAir	2002	39 A320, B737	Singapore Domestic Indonesia
	Citilink	2001	10 B737, A320	Indonesia Domestic
	Indonesia AirAsia	2000	17 A320	Indonesia Malaysia
	Lion Air	2000	79 B737	Singapore
Japan	Air Next	2005	14 B737	Japan Domestic
	Air Do	1998	10 B737-767	Japan Domestic
	Amakusa Airlines	2000	2 Bombardier Dash 8-183	Japan Domestic
	JAL Express	1998	28 B737-800	Japan Domestic
	Skymark Airlines	1998	24 Boeing 737-800	Tokyo Domestic
	Skynet	2007	12 B737	Japan Domestic

Table 7.3 Continued

Country	Low cost carrier	Year commenced operations	Fleet size (e.g. A320/330 or B737)	Destinations/region
Japan	Asia Airways	2007	3 Antonov An-12 1 Yakovlev Yak-40 1 Boeing 727	Tajikstan, India, China
	Star Flyer	2006	6 A320	Japan Domestic
South Korea	Air Busan	2008	3 Boeing 737-400 3 Boeing 737-500 2 Airbus A321-200	10 China, Japan, South Korea, Philippines, Taiwan
	Eastar Jet	2009	5 Boeing 737-700 1 Boeing 737-600	7 Bangkok, Japan, Seoul, Malaysia, China, Thailand
	Jeju Air	2006	8 Boeing B737-800	12 Hong Kong, Japan, Philippines, Thailand, Seoul, Busan, Cheongju
	Jin Air	2008	5 Boeing 737-800	7 South Korea, China, China SAR, Guam, Japan, Philippines, Thailand
	T'way Airlines	2005	4 Boeing 737-800	5 Taiwan, Japan, Cambodia, South Korea, Thailand
Malaysia	AirAsia	1996	56 Airbus A320-200	80+ Australia, Bangladesh, Brunei, Cambodia, New Zealand, Philippines, South Korea, Singapore
	AirAsia X	2007	9 Airbus A330-300 2 Airbus A340-300	14 Brunei, Burma, Cambodia, Indonesia, Laos, Vietnam, Thailand
	Firefly	2007	12 ATR 72-500	19 Indonesia, Malaysia, Singapore, Thailand
Philippines	Airphil Express	1996	11 Airbus A320-200 3 Bombardier Dash 8 Q300	26 Hong Kong, Philippines, Luzon, Mindanao, Visayas, Singapore
	Cebu Pacific	1996	10 Airbus A319-100 21 Airbus A320-200 8 ATR 72-500	52 China, Hong Kong, Japan, Macau, South Korea, Taiwan, Philippines, Vietnam

Table 7.3 Continued

Country	Low cost carrier	Year commenced operations	Fleet size (e.g. A320/330 or B737)	Destinations/region
Philippines	Spirit of Manila Airlines	2008 - currently not in operation	2 McDonnell Douglas MD-83 1 Boeing 737-300	5 Philippines, Taiwan, Palau, Macau
	Zest Airways	1996	1 Airbus A319-100 8 Airbus A320-200 4 Xian MA60	26 China, Taiwan, South Korea, Philippines
Singapore	Jetstar Asia Airways,	2004	16 Airbus A320-200 2 Airbus A330-200	21 Singapore, Hong Kong, New Zealand, Cambodia, China
	Tiger Airways	2005	21 Airbus A320-200	28 Singapore, Thailand, Singapore, Malaysia, Hong Kong, Philippines, Taiwan
	Valuair	2004	2 Airbus A320-200	5 Indonesia, Denpasar, Jakarta, Medan, Surabaya, Singapore
Sri Lanka	Mihin Lanka	2007	1 Airbus A320-232 1 Airbus A321-231	9 Bangladesh, India, Indonesia, Kuwait, Sri Lanka, United Arab Emirates
Thailand	Nok Air, Orient Thai Airlines, Thai AirAsia	2004	1 ATR 72-200, 10 Boeing 737-400 3 Boeing 737-800	30 Bangkok, Phuket, Mae sot, Trang, Koh Lanta, Koh Phangan
Vietnam	Jetstar Pacific/Qantas	2001	5 Boeing 737-400 5 Airbus A320-200	6 Vietnam, Da nang, Hanoi, Ho Chi Minh City
LCCs operating to Pacific destinations from Australia and New Zealand				
Fiji	Jetstar/Qantas Pacific Blue	2004	46 Airbus A320-200 10 Boeing 737-800	34 Australia, New Zealand, Fiji
Rarotonga	Pacific Blue/Virgin Blue	2004	10 Boeing 737-800	15 New Zealand, Australia, Pacific Islands
Vanuatu	Pacific Blue/Virgin Blue	2004	10 Boeing 737-800	15 New Zealand, Australia, Pacific Islands
Samoa	Pacific Blue/Virgin Blue	2004	10 Boeing 737-800	15 New Zealand, Australia, Pacific Islands
Tonga	Pacific Blue/Virgin Blue	2004	10 Boeing 737-800	15 New Zealand, Australia, Pacific Islands

Table 7.4 Visitors to selected low cost airline websites in Asia Pacific (February 2011 vs. February 2010)

Total audience Asia Pacific: Visitors aged 15+, home and work location*			
	Total unique visitors (000)		
	Feb 2010	*Feb 2011*	*% Change*
Total Internet: Total audience	472,864	545,183	15
Airasia.com	2,236	3,380	51
Tiger Airways	554	1,805	226
Jetstar.com	1,028	1,169	14
Cebupacificair.com	457	669	47
Goindigo.in	299	508	70
Skymark.co.jp	332	453	36
Spicejet.com	332	430	30
Fireflyz.com.my	143	269	89

Note: * Excludes visitation from public computers such as Internet cafes or access from mobile phones or PDAs.
Source: comScore Media Metrix 2011.

just in ASEAN', he said, referring to the ten-member Association of Southeast Asian Nations. India and China, which are already served by budget carriers from ASEAN cities, have a combined population of 2.5 billion. Rapid economic growth in ASEAN, China and India, coupled with falling airfares, means millions more can now fly rather than travel by land or sea compared to the pre-budget airline era. 'The exponential growth for low-cost carriers in Asia–Pacific is supported by emerging economies and their citizens' soaring income' (Lim 2011).

In the Asia Pacific region, LCAs have generated significant growth in online bookings since 2011 as travel offers and promotions have prompted more consumers – especially younger travellers – to consider LCC airlines (see Table 7.4). AirAsia grew its audience by 1.2 million visitors in the past year, which helped maintain its lead as the top-visited LCA site in the Asia Pacific region. Tiger Airways posted the strongest rate of growth, more than tripling its online traffic to 1.8 million visitors, while Malaysia's Firefly and India's IndiGo posted strong growth rates of 89 per cent and 70 per cent respectively (comScore 2011, see Media Metrix).

Case Study: Impact of Low Cost Carriers on Selected Pacific Islands' Economies

The air transport sector in the South Pacific continues to play a vital role in the economic growth of these sparsely populated islands scattered across great expanses of the South Pacific Ocean. Most airlines in the region are wholly or partially owned by governments and still operate on an outdated model, enjoying government subsidies and operating under bilateral ASAs. Deregulation and multilateral ASAs have opened up destinations such as the Cook Islands, Samoa, Tonga and Vanuatu for LCC services. LCCs have opened more frequencies, capacity and secondary routes at much lower fares than traditional airlines have been offering over the past 16 years (Kissling 2002 and Taumoepeau 2010a).

The first LCC to fly into the Pacific Islands was Kiwi International in 1994. It flew to Tonga and Samoa for a couple of years, but was undercapitalized and failed financially in 1996. Air New Zealand formed a LCC subsidiary, Freedom Air, in 1995, and competed with Kiwi International on regional flights, which quickly put Kiwi International out of business. The FreedomAir strategy for product and services, merged with the new restructured Air New Zealand in 2008. Jetstar, established in May 2004, is fully owned by Qantas, and its headquarters are in Melbourne, Australia. It is an integral part of the Qantas Group's two-brand strategy, operating in the leisure and value-based market. Jetstar started flying to Fiji in 2010 using A320s and has helped to boost tourism to Fiji from the large Australian market.

The LCC service to the Pacific, begun in 2005 by Pacific Blue (owned by Virgin Australia), flying seven Pacific routes from Australia and New Zealand, produced dramatic passenger growth and substantial economic benefits. It also resulted in considerable competitive pressure on fare levels, to the benefit of consumers. The Fiji Islands, Samoa, Tonga, and Vanuatu were the markets targeted by LCCs. Passenger traffic and visitor arrivals in all four target countries increased substantially (Asian Development Bank 2009 and Taumoepeau 2010b).

The LCCs' arrival and the associated economic benefits to the region have helped some Pacific islands to get back on track to achieve their Millennium Development Goals (MDGs) which were endorsed by the countries in 2002 (Pacific Islands Forum Secretariat 2010). Few countries other than Samoa and Tonga were dependent on remittances, but there was a general downturn in monies remitted from residents abroad. Other problems in the region such as natural disasters and the recent earthquakes in Samoa and Tonga slowed down the expected economic recovery of these countries. Fortunately, the tourism industry has recently picked up in some countries. In the South Pacific, tourism is the largest export sector for most countries and offers great opportunity for economic growth, employment and sustainable development (DeLacy 2009).

In the South Pacific region, Pacific Blue, a subsidiary of Virgin Australia, operates frequent services from New Zealand and Australia to the Pacific Islands. These new low fare air services opened up new markets for travelling to the islands and also increased spending by travellers in the South Pacific destinations visited.

Figure 7.5 shows corresponding recent improvements in the GDP of Fiji, attributed mainly to the arrival of LCAs. National tourism offices reported that they have witnessed an increase in the number of international arrivals from Australia and New Zealand. Local businesses also reported an increase in expenditure by

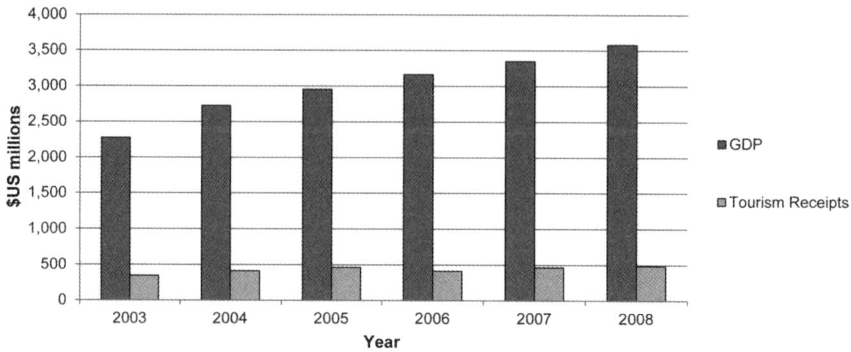

Figure 7.5 Fiji Summary of gross domestic product and tourism receipts 2003–2008

Source: Euromonitor International 2009.

travellers and this is explained by the low fares offered by the LCCS: travellers are spending the savings on shopping at their destinations; and returning residents spend more money on family and helping local families.

For years the government of Samoa was struggling financially to fund its national carrier, Polynesian Airlines, then in 2005 a joint venture was signed between the Government of Samoa and Virgin Blue. This new airline became

Figure 7.6 Virgin Samoa is a joint venture of the Samoan Government and Virgin Australia

Source: Michael Lück.

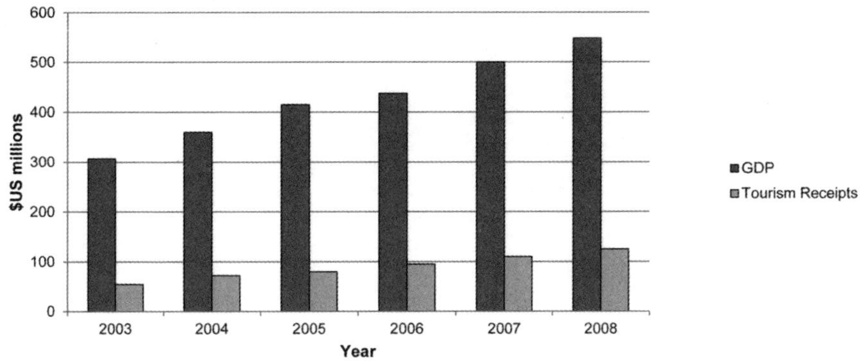

Figure 7.7 Samoa: Summary of gross domestic product and tourism receipts 2003–2008

Source: Euromonitor International 2009.

Polynesian Blue (it was renamed Virgin Samoa in 2012) and within the first 12 months the new airline, operating on a low cost model, was making a good profit and started to pay back dividends to the Samoa Government (see Figures 7.6 and 7.7). From October 2005 to February 2008, just over two years since Polynesian Blue was launched (sister LCC company Pacific Blue operates its international routes), holiday arrivals in Samoa from Australia and New Zealand increased by 46 per cent.

The number of holiday arrivals to Tonga from New Zealand and Australia increased by 54 per cent during the same period. Since Pacific Blue commenced flights to and from Vanuatu in 2004, the market has also grown significantly, with traveller numbers from Australia increasing by 52 per cent (see Table 7.5). Pacific

Table 7.5 Number of international air travellers to key LCC destinations (2005–2009)

Country	LCA start date	2005	2006	2007	2008	2009	% growth 2005–2009
Cook Is.	March 2005	88,405	92,328	97,316	94,776	100,592	14%
Fiji	Sept. 2004	545,145	548,589	539,881	585,031	542,186	*7%
Samoa	Oct. 2005	101,807	115,882	122,352	122,222	128,830	27%
Tonga	Oct. 2005	42,000	39,451	46,040	49,400	50,645	21%
Vanuatu	Sept. 2004	62,123	67,787	81,345	90,657	100,000 (est.)	46%

Note: * 2008 figure used. Tourism arrivals to Fiji in 2009 decreased due mainly to some political problems that cut down the number of travellers from New Zealand and Australia, but the situation has since normalized.

Source: Taumoepeau 2010b.

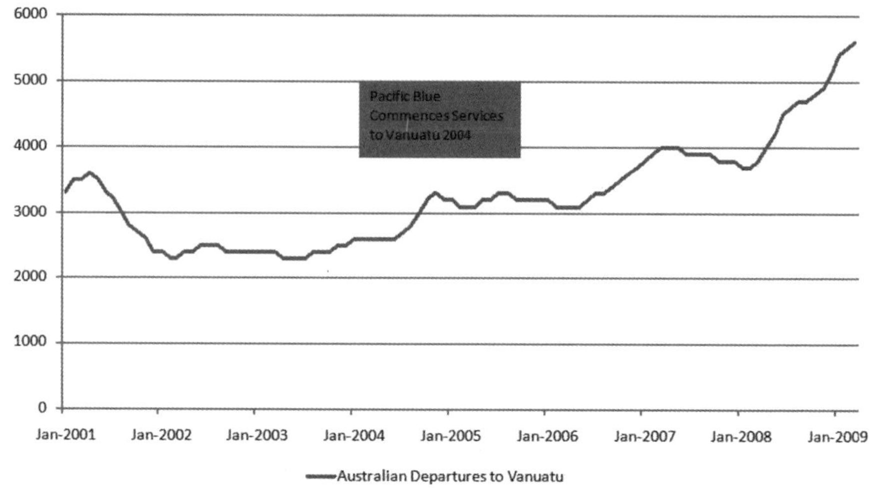

Figure 7.8 Australian Departures to Vanuatu
Source: Australia Bureau of Statistics 2009.

Blue, with services between Auckland and Rarotonga, Cook Islands, has two direct flights a week from New Zealand to the Cook Islands, and increased tourist traffic by nearly 28 per cent in just over two years from its launch in October 2005. Pacific Blue has also grown its Fiji services from 12 to 24 flights a week, which has had a major impact on the local economy; and with over 161,000 visitors flying Pacific Blue to Nandi since it started the service, the economic impact has been assessed to be in excess of Fiji $250 million. Table 7.5 shows that selected LCC destinations such as the Cook Islands, Fiji, Samoa, Tonga and Vanuatu showed double digit growth after the introduction of the LCC services in late 2004 and during 2005.

Figures extracted from the Euromonitor International 2009 publications showed corresponding increase in tourism receipts and GDP during the same period in some of the selected destinations, notably Fiji, Samoa and Vanuatu. Figures from the Australian Bureau of Statistics (2009) and Vanuatu National Statistics Office (2009) show the trend in short-term departures to Vanuatu from Australia in 2004 (see Figure 7.8).

Future Trends in the Asia Pacific Region

Proliferation of the newly established LCCs as subsidiaries of established national airlines is a strategic initiative to help grow the stagnant revenue of the traditional carriers. The first six months of 2012 have seen steady traffic growth in the Asia–

Pacific region, with an increase in both city-pairs and seat capacity of 8 per cent over the same period in 2011. China in particular continues to be the growth engine for the region, with nearly 2.0 million additional seats available for June 2013 over June 2012 in the domestic market alone. Air China, China Eastern and China Southern have also driven a strong growth in international service of 14 per cent year on year increase in international frequencies and 15 per cent increase in international seat capacity.

The short-term challenges to the Japanese market are reflected in a number of short-term service reductions by both Japan Airlines (JAL) and All Nippon Airways (ANA) that drive the minor reductions in seat capacity in the Japanese market in June 2011.

The Indian domestic market continues to record strong growth, with an 18 per cent increase in seat capacity. LCCs are driving much of this domestic growth, and they now have a 60 per cent market share of the capacity within India. However, some of the national airlines in the region such as Japan Airlines, Thai Airways and Singapore Airlines have incurred some losses and are undergoing some cost control measures. Against this background, the LCCs of the region have used the global economic crisis as an opportunity to gain market share and consolidate their positions. Most of the Asian LCCs are using a fleet of single type fuel-efficient Airbus 320 or Boeing B737, allowing them to save on spare parts, maintenance and training. LCCs develop and implement creative ways to raise non ticket-related income (ancillary revenues) by unbundling products and services and giving passengers the option to purchase onboard meals, insurance policies, baggage, loyalty programmes, hotels and related services (AirAsia).

When Malaysia's LCC AirAsia started in 2001, only 6 per cent of Malaysians had flown before. Under a 'Now everyone can fly' marketing slogan, this budget carrier has frequently offered ticket prices lower than some bus fares. The LCCs have certainly changed the way people travel; less affluent people who cannot afford to pay for FSCs can now afford to fly. The recent deregulation of Southeast Asia's skies has opened the industry to genuine price competition after decades of monopolistic collusion among national flag carriers. The Malaysia–Singapore route, for example, was only recently opened to competition after Singapore Airlines and Malaysia Airlines dominated the route for over 35 years. Duopolistic behaviour resulted in one of the most expensive routes in the world for a 55-minute flight, with ticket prices routinely over $US400. LCCs now offer fares for a quarter of that amount and with much higher frequency. AirAsia travels between Kuala Lumpur and Singapore around nine times per day (Wijaya 2010).

Southeast Asia's Open Sky Agreement, which will come into full effect by 2015, is expected to further benefit the region's LCCs. The agreement will allow regional air carriers to operate unlimited flights to all ten ASEAN member countries and promises to boost intra-regional tourism, trade and greater economic growth among member countries: Brunei, Cambodia, Indonesia, Laos, Malaysia, Myanmar, the Philippines, Singapore, Thailand and Vietnam.

A recent report from the Sydney-based Centre for Asia Pacific Aviation predicts future industry consolidation among smaller players, many of which it predicts will be forced to merge or close as competition heats up (CAPA 2011).

In the past, ANA and JAL tried to meet the demand of all types of passengers with a single traditional airline business approach. Having launched the new airline, they are now going to concentrate their management resources on specific market segments. They are also seeking to realize the efficient management and the market adaptation simultaneously. To this end, it is essential to have a new aviation business model and to differentiate its playing field from that of legacy carriers (Nomura Holdings 2011).

As explained earlier, LCCs in Asia operate either as a subsidiary of an established national airline such as AirAsia (Malaysia Airlines), Jetstar (Qantas), Tiger Airways (Singapore Airline) and Nok Air (Thai Airways), or as a new LCC identity such as Air Deccan, formed in 2003 in India, and Spring Airline, formed in 2005 in China. The Pacific does not have any LCC based in the region (in contrast to Asia or Australia). However, Pacific Blue, a subsidiary of LCC Virgin Blue, and Jetstar service various destinations in the South Pacific from Australia and New Zealand. National airlines such as Air New Zealand, Air Pacific, Air Vanuatu, Aircalin and Air Niugini have yet to establish a separate LCC subsidiary, as do some Asian FSCs. This is due mainly to the fact that these national Pacific airlines are still majority government owned and shareholders are still reluctant to take on these new ventures. Traditionally national airlines of the South Pacific are in business for a variety of reasons, such as acting as a public utility, linking the capital city with the other islands. Their operations are still based on bilateral rights agreements and they are not willing to go into the LCC model, because traffic is light and destinations are isolated with varying airfield standards, some not fit for any of the LCC models of using one type aircraft such as the B737 or A320.

Roughly half of the future narrowbody deliveries are destined for the region's LCCs. This means their fleet share will rise to approximately 33 per cent of the total by 2015. This will equate to a combined Asia Pacific narrowbody LCC fleet of about 985 aircraft in 2015, from 465 in 2010. But China's 'big three' state owned carriers are heavily represented in the narrowbody order books for Airbus and Boeing. Excluding mainland China (where home-grown LCCs are not sanctioned and most of those aircraft are destined to cover domestic Chinese growth), the Asia Pacific LCC expansion story is even more dramatic.

Singapore Airlines (SIA) announced the launch of another new budget airline (Scoot) because its mainline operation is not growing. In 2010, the airline carried 16.6 million passengers, compared with 19 million in 2007–2008. In the same period, Singapore's Changi Airport saw its passenger traffic growing from 36 million to 44 million. Given the more open-air regulations in Singapore vis-à-vis other Asian locations, SIA faces more competitive pressures than some of its peers. For example, LCCs have a market share of 27 per cent in Singapore compared to less than 5 per cent in Hong Kong (Nomura Holdings 2010).

This is a similar 'dual brand' strategy to that pioneered by Qantas and Jetstar. The Jetstar subsidiary generated one-third of the Qantas Group's annual profit in 2011. Qantas expansion plans would centre on Jetstar, as opportunities for the mainline Qantas brand were limited. Singapore Airlines already has a 33 per cent stake in budget carrier Tiger Airways, which it set up in 2003. Tiger operates a fleet of 26 A320s on regional routes from Singapore and within Australia.

Another Asian FSC planning to launch a LCC is Thai Airways (THAI), which in 2011 announced a joint venture with Tiger Airways to establish Thai Tiger. Thai Tiger will offer short haul services out of Bangkok's Suvarnabhumi International Airport and planned destinations include Chiang Mai, Phuket, Penang, Kuala Lumpur, Macau, Shenzhen and Chennai. THAI also has a 49 per cent stake in the regional LCC Nok Air and plan to set up another LCC carrier, similar to the regional operations of Singapore Airlines (Silk Air) and Cathay Pacific (Dragonair).

Meanwhile, ANA, in partnership with Hong Kong investors First Eastern and Japan's Innovation Network Corporation, set up a low cost subsidiary called Peach, which started flying from Osaka's Kansai International Airport to Fukuoka, Sapporo and Seoul in 2012. Introductory fares are 50 per cent lower than those offered now by ANA and JAL. With Peach, ANA aims to stimulate additional demand for air services within Japan, whose economy is barely growing, by pricing tickets at levels that make them competitive with high-speed rail.

Japan Airlines (JAL) and oneworld partner Qantas are reportedly considering starting up a LCC for domestic flights in order to capitalize on the know-how of Qantas's subsidiary Jetstar. Jetstar and Jetstar Asia already fly long haul low cost routes from Australia and Singapore to Japan, and the carrier has been emphasizing its intention to form other joint ventures in Asia in order to expand its footprint in the world's fastest growing aviation market. Qantas also partly owns a stake in Vietnam's second-largest airline, Jetstar Pacific.

Thus the trend for the LCC sector to dominate the skies of the Asia–Pacific region continues. Depending on individual national aviation policies, the nature of the market, local economies and competition from the traditional carriers, the LCCs have adopted local structures, services and routes conducive to their growth strategies. With the lowering of airfares, making flying very affordable for a growing majority of the population, millions more will use the LCC services to travel for business and for leisure purposes, thus further developing the economies of these countries.

References

Airline Business (2011). Airline Industry Guide 2010/11, May. Sutton, UK: FlightGlobal.

Airlinetrends.com (2001). *Asia's full-service airlines go low-cost*, 31 May. Available at: http://www.airlinetrends.com/2011/05/31/asian-full-service-airlines-go-low-cost/ [accessed: 30 November 2011].

Asian Development Bank (2007). *Oceanic Voyages: Aviation in the Pacific* (Pacific Studies Series). Manila: ADB Publications. Available at: http://www.adb.org/sites/default/files/pub/2007/aviation-in-the-pacific.pdf [accessed: 15 May 2012].

Asian Development Bank (2009). *Pacific Economic Monitor*, August. Available at: http://www.adb.org/sites/default/files/pub/2009/pem-aug09.pdf [accessed: 13 January 2011].

Asian Development Bank (2012). *Annual Report 2011*. Mandaluyong City, Philippines: ADB.

Australian Bureau of Statistics (2009). File: ABS-3401.0-Overseas Arrivals Departures Australia-Shortterm Movement Resident Departures Selected Destinations-Seasonally Adjusted-Persons-Vanuatu-Short Term Less Than One Year Residents Departing. Available at: http://commons.wikimedia.org/wiki/File:ABS-3401.0-OverseasArrivalsDeparturesAustralia-ShorttermMovementResidentDepartures_SelectedDestinations-SeasonallyAdjusted-Persons-Vanuatu-ShortTermLessThanOneYearResidentsDeparting-A2263649C.svg [accessed: 15 July 2012].

CAPA (2004). *Asia Pacific Aviation Outlook 2004: 2004 – A Year of Massive Opportunity for Aviation & Tourism*. Sydney, Australia: Centre for Asia Pacific Aviation (CAPA). Available at: http://www.centreforaviation.com/reports/files/16/Asia%20Pacific%20Outlook%202004.pdf [accessed: 12 May 2012].

CAPA (2011). *Low Cost Share of Capacity 2010*. Sydney, Australia: Centre for Asia Pacific Aviation (CAPA).

CAPA (2012). *Low Cost Carriers (LCCs): LCC Capacity Share (%) of Total Seats: 2001–2012*, Sydney, Australia: Centre for Asia Pacific Aviation (CAPA). Available at: http://www.centreforaviation.com/profiles/hot-issues/low-cost-carriers-lccs#lcc [accessed: 6 June 2012].

comScore (2011). *Low-Cost Airlines Take Off Online in Asia Pacific*, 11 April. Singapore: comScore Inc. Available at: http://www.comscore.com/Press_Events/Press_Releases/2011/4/Low-Cost_Airlines_Take_Off_Online_in_Asia_Pacific [accessed: 12 June 2011].

DeLacy, T. (2009). *Pacific Tourism Adaptation to Climate Change Risks*. Presentation to the UNDP Conference on Climate Change in the Pacific: Impact on Local Communities, Apia, Samoa, 1 August 2009.

Doganis, R. (2006). *The Airline Business* (2nd ed.). New York: Routledge.

Euromonitor International (2009). *The World Economic Factbook 2009* (16th ed.). London: Euromonitor International Ltd.

Forsyth, P. and King, J. (1996). Cooperation, competition, and financial performance in South Pacific aviation, in Hufbauer, G. and Findlay, C. (eds), *Flying High: Liberalising Aviation in the Asia Pacific*. Washington, DC: Institute for International Economics, November.

Graham, A., Papatheodorou, A. and Forsyth, P. (2010). *Aviation and Tourism: Aviation and Tourism: Implications for Leisure Travel*. Farnham, UK: Ashgate.

IATA (2010). *AGM Report*, June. Geneva: International Air Transport Association.

ICAO (2009). Annual Reports of the Council. Bangkok: ICAO. Available at: http://www.icao.int/publications/pages/annual.reports.aspx [accessed 23 March 2013].

Kissling, C. (1998). Liberal aviation agreements – New Zealand. *Journal of Air Transport Management*, 4(3), 177–80.

Kissling, C. (2002). *Transport and Communications for Pacific Microstates: Issues in Organisation and Management*. Suva, Fiji: Institute of Pacific Studies of the University of the South Pacific.

Lim, P. (2011). *Asian Budget Carriers Spread Wings as Demand Surges*. Asia Business Aviation (AFP). 16 June. Available at: http://www.google.com/hostednews/afp/article/ALeqM5ir0eeybLFCh_ieZVPdnK4Yh-zKeA?docId=CNG.5eb0f8d3d40d16ce25bfc129328515e1.7d1 [accessed: 15 November 2011].

Maidment, R. and Mackerras, C. (eds) (1998). *Culture and Society in the Asia-Pacific*. London: Routledge/Open University.

National Bureau of Asian Research (2011). *Politics in Asia*. Seattle, WA: NBR. Available at: http://nbr.org/research/theme.aspx?id=bc960f0d-7201-4549-ae3a-e6daff822a70 [accessed: 15 July 2012].

Nomura Holdings (2010). *Nomura Investments Report 2010*. Tokyo: Nomura Holdings Inc.

Nomura Holdings (2011). *SEC Filings*. Available at: http://www.nomuraholdings.com/investor/library/sec/ [accessed: 30 November 2011].

O'Connell, J.F. and Williams, G. (2006). Transformation of India's domestic airlines: A case study of Indian Airlines, Jet Airways, Air Sahara and Air Deccan. *Journal of Air Transport Management*, 12, 358–74.

Pacific Forum (2004). Pacific Forum Report on Regional Transport Services, delivered at the Forum Leaders Meeting, Apia, Samoa, 6 August. Available at: http://www.forumsec.org.fj/resources [accessed 23 March 2013].

Pacific Islands Forum Secretariat (2010). *Annual Report 2010*. Suva, Fiji: Pacific Islands Forum. Available at: http://www.forumsec.org/resources/uploads/attachments/documents/PT&I%202010%20Annual%20Report%20FINAL.pdf [accessed: 15 July 2012].

Papatheodorou, A. (2010). The impact of civil aviation regimes on leisure travel, in Graham, A., Papatheodorou, A. and Forsyth, O. (eds), *Aviation and Tourism: Implications for Leisure Travel*. Aldershot, UK: Ashgate, 49–58.

PATA (2011). *Special Report*. Bangkok: Pacific Asia Travel Association, 12 April.

Taumoepeau, S. (2007). *A Blueprint for the Economic Sustainability of the Small National Airlines of the South Pacific*, Unpublished DBA thesis, University of the Sunshine Coast, Queensland, Australia.

Taumoepeau, S. (2009). *South Pacific Aviation*, Unpublished DBA thesis, University of the Sunshine Coast, Queensland, Australia.

Taumoepeau, S. (2010a). South Pacific, in Graham, A., Papatheodorou, A. and Forsyth, O. (eds), *Aviation and Tourism: Implications for Leisure Travel*. Aldershot, UK: Ashgate, 323–31.

Taumoepeau, S. (2010b). *Impact of Low Cost Airlines on Pacific Island Economies*, New Zealand Tourism and Hospitality Research Conference (NZTHRC), Auckland, New Zealand, 24–26 November.

Turner, L.W. and Witt, S.F. (2011). *Asia Pacific Regional Tourism Forecasts 20011–2013*. Bangkok: Pacific Asia Travel Association.

Vanuatu National Statistics Office (2009). Vanuatu National Statistics, *Visitors Arrivals by Country of Residence 2007–2009*. Port Vila, Vanuatu: Private Mail Bag 9019.

Wijaya, M. (2010). New heights for Asia's budget carriers, 17 February. *Asian Times*. Available at: http://www.atimes.com/atimes/Southeast_Asia/LB17Ae01.html [accessed: 30 November 2011].

World Bank (2011). *The World Bank Annual Report 2011: Year in Review*. Washington, DC: World Bank.

Chapter 8
Low Cost Carriers in India

Tarun Shukla

Introduction

India, which in 2011 celebrated 100 years since its first flight, was a late entrant to the world of low cost carriers (LCCs).The first low cost flight took off in 2003 when Air Deccan was launched, creating a curious storm in the skies and changing the dynamics of air travel in India. Today nearly 70 per cent of Indian passengers fly on economy flights, of which about 50 per cent are exclusively dominated by four dedicated low cost airlines (LCAs). With only 60 million domestic passengers flying, out of a population of 1.2 billion, the future for discount carriers holds immense potential. This chapter discusses how prepared these LCCs are to capture this growth and, more importantly, to figure out if they have what it takes to dominate the Middle Eastern, central Asian and Southeast Asian skies which, so far, have been the monopoly of a few.

India: Political, Social and Economic Factors

India dominates the South Asian subcontinent. To the west it borders the Arabian Sea and to the east the Bay of Bengal; and it shares land borders with Bangladesh, Myanmar, China and Pakistan. India is an emerging economy, which, with China, Russia, Mexico and Brazil, has undergone unprecedented levels of economic expansion in recent years. The country has a cost-effective and labour-intensive economy that benefits from outsourcing, while having a strong manufacturing and export-oriented industrial framework.

India's GDP has been growing at 8.5 per cent a year on average in the last five years, though it is expected to grow at about 7 per cent in 2011–2012, down partly because of the global economic environment (see Table 8.1 and *Economic Times* 2011a). Moody's adds that:

> expectation is that supported by current levels of savings and investment, growth will revive over the medium-term, thanks to continued productivity enhancements in the private sector, increased infrastructure investment boosting potential output, policies to alleviate poverty and income inequality that will support domestic demand growth, and demographic trends raising the working age population while keeping the dependency ratio low (*Economic Times* 2011a).

Table 8.1 Asian economies' growth rates

Country	Total population (m)	GDP per capita at current prices (US$)	2012 GDP growth forecasts*
China	1,341.4	4,382	9 %
India	1,215.9	1,265	8 %
Indonesia	234.4	3,015	7 %
Philippines	94.0	2,007	6 %
Vietnam	87.2	1,174	7 %
Thailand	66.9	4,992	5 %
Malaysia	61.2	702	6 %
Australia	28.3	8,423	6 %
Cambodia	22.2	55,590	5 %
Laos	15.0	814	6 %
Singapore	5.2	43,117	6 %
Brunei	0.4	31,239	3 %

Source: Webb 2011.

Overall India is being driven by young demographics, strong domestic consumption-led growth, a growing middle class, increasing productivity, strong entrepreneurial culture, a more transparent and accountable government, big businesses and markets, politics of aspiration and inclusion, and growing urbanization.

At present 1 in 200 Indians flies just once a year. Ireena Vittal, a retail specialist at McKinsey, says India's 1.2 billion people can be divided into roughly 250 million households (Pilling, Hille and Kazmin 2011). Of those, 100 million live in poverty and have little realistic prospect of attaining middleclass status. Just 2 million households enjoy the same standard of living as their rich counterparts in the USA or Europe. There are 14–15 million households with an annual income of US$7,000–$10,000 – a number set to explode to 40 million households or 200 million people within five years. Even if India's economy grows at an annual 7.3 per cent – below the current 8.5 per cent – McKinsey estimates that by 2025 it will have a middle class of 580 million people (Pilling, Hille and Kazmin 2011).

Given that only 60 million people flew within India in 2011, airlines are banking on the growth potential of the domestic market. Anticipating significant growth in traffic, most Indian carriers have placed orders to augment their aircraft fleets. India is expected to add around 370 aircraft worth 150,000 Indian rupees (INR) crore (US$33.3 billion) in their fleet by 2016–2017. Fleet expansion at this scale would require airlines to explore multiple funding options including capital markets, long term borrowings and leasing (Civil Aviation Ministry 2011).

The Aviation Industry in India

India is sandwiched between the heavyweight airport mega-hubs around it. This includes the Middle East (Dubai, Doha and Abu Dhabi) and the Asian quartet (Bangkok, Hong Kong, Kuala Lumpur and Singapore). These hubs have seen evolution over the past 20–30 years fed largely on traffic from India and growth of Asian Tiger economies.

Aviation contributed 1.5 per cent of India's GDP in 2009, supporting 9.95 million jobs (Civil Aviation Ministry 2011). However, air travel penetration in India at 0.04 trips per capita per annum is much less than in the USA, with an average of two air trips per capita per annum.

If GDP grows at 7 per cent, India's aviation market growth will continue to accelerate at double digit rates over the coming decade, according to most aviation analysts, airline owners and March 2012 projections made by the Indian Government Ministry of Finance.

Average growth of passenger traffic is expected at an annual rate of 12 per cent between 2011–2012 and 2016–2017. The domestic passenger volume is expected to grow to approximately 209 million by 2016–2017, from 106 million in 2011–2012. Similarly, international passenger volume is estimated to grow at an average annual rate of 8 per cent during the same period to reach 60 million passengers by 2016–2017, from 38 million in 2011–2012 (Civil Aviation Ministry 2011).

From 500 departures per week in the country in 1994 (before deregulation), to nearly 15,000 daily departures, and with 87 foreign airlines flying to and from India, together with five Indian carriers to and from 40 countries, the landscape of aviation has changed. India's domestic aviation market expansion has been the strongest in the world, tripling in the past five years to become the ninth largest aviation market in the world. It is expected to be among the top five over the next ten years (*The Hindu* 2011).

Seven passenger airlines competed for passenger in the home market in 2012. These include state-run Air India, Jet Airways, IndiGo, Kingfisher Airlines, SpiceJet, JetLite and GoAir (see Figure 8.1 and Table 8.2).

Fleets of Indian carriers are dominated by Airbus and Boeing aircraft. Air India uses a mixed fleet of 135 aircraft fleet and has arguably the most expensive aircraft in its fleet. These include Boeing B777 LR/ER to connect with North America with 16-hour non-stop flights, Airbus A330 for Europe, Airbus A320 for domestic flights, and Boeing B737 for flights to the Middle East and Southeast Asia, and Bombardier CRJs and ATRs to connect shorthaul destinations. Some destinations in Europe and South East Asia are also connected with Airbus A330s and Boeing 777s.

Air India is also expected to receive 27 Boeing B787s, as one of its first customers, in 2012, and is on standby to join the Star Alliance. Jet Airways also uses Boeing B777LRs and the Airbus A330 for its international flights and operates from a European hub in Brussels. Its Boeing B737s are used for domestic flights and ATRs for smaller cities. Kingfisher has a fleet of Airbus A320, A330 and ATR

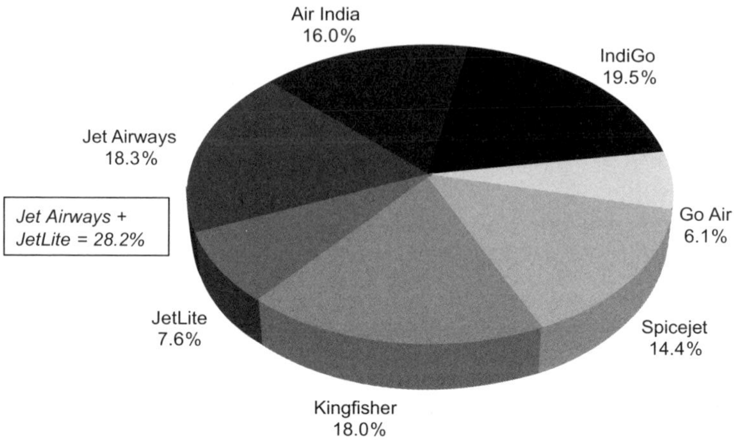

Figure 8.1 The airline market share in India (2012)
Source: Directorate General of Civil Aviation (2012).

aircraft. It was expected to join the oneworld alliance in February 2012 but that has been put in abeyance as the carrier has gone into a deep financial crisis, reducing its operational fleet to 20 aircraft of the 61 it has currently registered. The decision was taken – oneworld says – to allow the airline to meet is financial targets and a new date will be announced then.

LCAs operate both Boeing and Airbus fleets. IndiGo uses only A320s with a single class configuration of 180 seats, as does GoAir, while JetLite (now JetKonnect) and SpiceJet operate their services with Boeing B737s.

The airline growth has been helped by infrastructure growth on the ground. Passenger handling capacity at the country's 125 airports has risen from 72 million to 233 million in the last five years. About US$7 billion has been invested in modernization of the country's six biggest major airports, in addition to the modernization of 24 non-metro airports. There are at least 80 airports with scheduled flights.

In 2005 the Indian government announced a US$10 billion airport modernization plan for the period through to 2010. Under this plan, Delhi and Mumbai Airports, which command over 60 per cent of India's traffic, have been modernized by private operators, while new airports were built at Bangalore and Hyderabad (see Table 8.3). Kolkata and Chennai airports were modernized by the government-owned Airports Authority of India (AAI), which also upgraded about three dozen state capital airports. Terminal Three (T3) at Delhi International Airport has been constructed at an estimated cost of US$2.8 billion and is the flagship of the Indian airport system which has the potential to be the first hub in the country. With the completion of Phase I work of Indira Gandhi International Airport, a new integrated T3 has become operational, with a 34 million passenger handling capacity per annum (covering both international and domestic passengers), 168 check-in counters, 24 remote check-in counters, modern 5-level in-line baggage

Table 8.2 Key indicators of the top international Indian airlines (September 2010 to September 2011)

Airline	International seats	% seats operated by Indian airlines	International ASKs	% ASKs operated by Indian airlines	International frequency	% frequency operated by Indian airlines
Air India	140,675	38.2%	620,348,773	46.6%	658	33.5%
Jet Airways	113,600	30.9%	402,204,050	30.2%	680	34.6%
Air India Express	67,284	18.3%	172,291,833	13.0%	356	18.1%
Kingfisher Airlines	31,894	8.7%	110,051,434	8.3%	194	9.9%
SpiceJet	8,064	2.2%	6,131,328	0.5%	42	2.1%
IndiGo	6,480	1.8%	18,939,240	1.4%	36	1.8%
Total	367,997	100.0%	1,329,966,658	100.0%	1,966	100.0%

Source: CAPA 2011a.

Table 8.3 Total passenger numbers handled in 2010

Airport	Passenger Numbers
New Delhi	29.9 million
Mumbai	29.1 million
Chennai	12.0 million
Bangalore	11.6 million
Kolkata	9.6 million
Hyderabad	7.6 million
Cochin	4.3 million
Ahmedabad	4.0 million
Goa	3.1 million
Pune	2.8 million

Source: CAPA India and SITA 2012.

system, 98 immigration counters, 78 air bridges, multi-level car parking (4,300 capacity), 3,000 CCTVs, and 352 screening machines. Over the next ten years the airport sector is expected to see a further investment of US$20 billion, which will consist of upgrades to existing airports, new airports, and provisioning for second airports at each of these metropolitan areas. The Delhi–Mumbai route is the world's 10th busiest airline route in terms of number of flights between the two cities (*The Economist* online 2012).

Indian airlines will need to buy around 1,320 new airplanes to meet the demand of an expanding aviation sector. Boeing forecasts a US$150 billion market for passenger airplanes in India over the next 20 years. For example, LCA IndiGo placed the single biggest order in Airbus commercial aircraft history in January 2011, for 180 Airbus A320 aircraft worth $US15 billion, and is expected to grow the fastest. It is already a 50 aircraft airline and controls just under 20 per cent of the market. Its total order book currently stands at 240 aircraft, which means on average it will add 20 aircraft annually over the next ten years, many of which will be returned as leases expire.

Regulatory Environment

Airfares in India are fully transparent to the public and travellers are able to select the lowest price option because of the Internet and the round-the-clock search facility. India has over 100 million Internet users and this number is growing (*Economic Times* 2011b). The advent of LCCs has meant that airfares for the Delhi–Mumbai sector, which were sold for 12,000INR (US$250) one way in 2001, are now being sold for one-third, at 4,000INR (US$78) on average, and can go up depending how late the passenger books the ticket. Severe competition means that the airlines are drastically reducing fares, although their costs may not be adapted to such low fares. Consequently, all Indian airlines with about 450 aircraft have a combined debt of US$20 billion, and the airline sector has seen a profitless growth between 2000 and 2010.

The big carriers Air India, Jet Airways and Kingfisher are the airlines with the largest share of this debt. The latter is on the verge of bankruptcy as lessors have threatened to repossess the aircraft, banks have refused loans, and losses are mounting. At the time of writing, 44 out of Kingfisher's 64 aircraft are in operation while the rest are on the ground due to a lack of spares, and some are being cannibalized (Sinha 2012).

Airline regulation in India suffers from both a lack of clearly laid down civil aviation policy and also an old-school belief that aviation is for the rich in a country where many live on US$1 a day. The Civil Aviation Ministry decides most of the issues governing the airlines. The regulator, the Directorate General of Civil Aviation (DGCA), India's equivalent to the FAA, has laid down strict rules but it comes under the umbrella of the aviation ministry and is not completely independent. A regulator has recently been established to regulate airport tariffs, given that private airports now dominate the landscape, unlike in Western countries, where they are run mostly by local authorities. Besides tough regulations, airlines are dragged down by the lack of a clearly laid out civil aviation policy and a very high cost environment, which is very soon going to include the high costs airports. The newly built airports are gearing up to pass their costs of billion dollar investments on to the passengers and airlines. An indication of the negative impact of the policy is that India's airlines have accumulated a combined US$8.5 billion in losses over the last six years (CAPA 2011b).

Some policy and regulatory issues that directly affect LCCs are:

1. *Airports*: India does not have lowcost airports or secondary airports that would allow LCCs to use them for a lower fee, while main city centre airports charge premium fees for FSCs. This means the costs of airport usage are the same for the LCCs and FSCs.
2. *Taxation*: The taxes in India on jet fuel are high, which means the cost of fuel to an airline's operations is as much as 40 per cent. Ironically the same fuel is cheaper for international flights, being operated by local and foreign airlines.
3. *Route dispersal guidelines*: Regulations require airlines to fly at least 10 per cent of their services to remote cities, which is uneconomical due to high costs.
4. *Five Year/20 Aircraft Rule*: Regulation prevents Indian carriers from operating overseas until they have completed five years of domestic operations and operate a fleet of at least 20 aircraft; yet even a foreign start-up airline is allowed entry to connect Indian cities with international flights.
5. *DGCA*: The regulator is under-resourced with only 130 staff for the country's growing aviation sector, which employs over 100,000 people.
6. *Airspace*: Civilian and military airspace is clearly allocated. However, most of the airspace is dominated by the military, and while some progress has been made with the dual use of existing airspace, this move has been limited to some regions only, not nationally. Restricted airspace in some cases means longer routes for commercial flights, so more time and, subsequently, a waste of fuel. Some airports managed by the military, such as Pune, have limited airport slots for civilian flights.

Several of these issues are known but they remain a work in progress even as giant leaps are being made in infrastructure development such as modern airports and satellite navigation systems.

Government policy on bilateral international flying rights, which LCCs desperately need now, also became controlled in 2011 after a wave of liberalization in the years from 2003–2010. This is partly because foreign airlines have added several hundred flights into the country. Emirates is one of the biggest carriers, operating more flights into India than even the homegrown Kingfisher Airlines.

India's liberal access is demonstrated by the fact that carriers such as Emirates operate 184 weekly frequencies, while Air Arabia operates to 13 destinations across the country. International air services agreements will increasingly be driven by India's strategic, economic and political interests, not those of individual airline companies. For all of 2011 there was a complete blockade by the government on issuing fresh bilateral agreements to foreign carriers and domestic private carriers. That is because the government auditor raised serious questions about the way rights were granted to foreign airlines which led to the financial breakdown of flag carrier Air India (Shukla 2011).

Consequently, carriers like Lufthansa, Emirates and Singapore Airlines have not been given permission to operate Airbus A380s into India. India follows an open skies policy for all cargo services, and has an open skies agreement for passenger airlines with the USA, under which designated US airlines are permitted to operate unlimited air services to/through any point in India via any intermediate point, to any point beyond and vice versa, utilizing any aircraft type, with full traffic rights.

History of Low Cost Carriers in India

In 2003, the ex-army officer turned farmer turned entrepreneur Captain Gorur Ramaswamy Iyengar Gopinath, or G.R. Gopinath, decided to do something out of the box and create an airline that would provide cheap fares so more people could take to the skies. The result was the launch of Air Deccan from the IT city of Bangalore in August 2003, India's first LCA.

The launch of Air Deccan, which was to be a 'common man's airline', was marred by a heavy cloud of suspicion over safety and quality standards of such a LCA. It used a turboprop ATR aircraft to start its inaugural flight, which caught fire, creating embarrassment for Gopinath and attracting negative press coverage. However, this was, as time would tell, hardly a roadblock that would stop Gopinath. Air Deccan grew exponentially and changed Indian travellers' perceptions of airlines (Gopinath 2010).

Air Deccan pioneered the sale of food onboard, tried to sell most of its tickets through its website and announced early booking sales for as low as 1INR to attract first-time travellers. After taxes and fees some of these 1INR (US$0.018067) tickets could cost on average US$20. Many travellers who would only use railways would now be able to afford the joy of flying. India's former aviation minister, Praful Patel, once said that thanks to LCCs like Air Deccan, India's airports started looking like railway stations, which is a reference to the surprising growth in the sector.

Air Deccan had a dream run over the next few years. Passenger growth jumped to 42 per cent during 2006–2007, compared with 29.6 per cent in 2005–2006. The airline was growing at breakneck speed. By 2007, it was India's second largest airline with 40 aircraft, including Airbus A320s, flying to 60 destinations and creating a threat to other airlines. Cities that never had air services by the country's state owned Air India were now connected and heavily dependent on Air Deccan's daily flights.

'My whole dream was to build this airline and not to make money for myself. I could have sold the business and retired on a fat compensation, and had that happened the entire history of Indian aviation might then have been different, as would the course of my life', Gopinath wrote in *Simply Flying*, a book which was released after he sold the airline to liquor tycoon Vijay Mallya in 2007 after funding problems.

Between 2003 and 2007, during Air Deccan's successful years, several LCCs entered the market. In 2005, SpiceJet and GoAir launched low cost operations in India and they were closely followed by IndiGo's launch in 2006. In 2007, Jet Airways took over Air Sahara and relaunched it as a LCA under the JetLite brand (in 2012 JetLite was again renamed JetKonnect).

The launch of Air Deccan coincided with a period of economic boom in a country where the need for travel was growing, buoyed by an expanding middle class that wanted to fly at affordable prices.

Sensing this need, Mallya launched Kingfisher Airlines three years after Air Deccan as a premium carrier. He bought Air Deccan for about US$200 million, merging Air Deccan with Kingfisher and taking over a combined domestic passenger market share of 29 per cent in 2007.

Mallya, some analysts feel, overpaid for Air Deccan because he was in a hurry. He wanted to fly Kingfisher internationally, for which he needed to complete at least five years on the Indian domestic market. By buying Air Deccan and merging it with Kingfisher, he overcame this requirement and started his international operations soon after the merger.

Kingfisher continued to operate all-economy flights on former Air Deccan aircraft and full-service flights with its own aircraft. Today, the Indian domestic market has at least three airline business models in operation. First, there are the 'low cost' airlines (LCAs), like IndiGo, GoAir and SpiceJet, which offer a no-frills product at prices lower than those of full-service carriers (FSCs). These are closely followed by the full-service airlines (FSAs), such as Jet Airways, which have introduced all-economy-class products (like JetKonnect), which do not have significantly lower costs but are a stripped-down product. About 30 per cent of Jet Airways fall into the third category, a dedicated FSC. Currently, Air India remains the only FSA product.

India's Low Cost Airlines

Compared to most of the LCCs across the world, Indian LCCs have developed a model of their own, offering an enhanced product in terms of value for money, but are yet to evolve a Southwest Airlines-style business. When compared to legacy carriers in the United States or Europe, Indian LCCs offer superior services right

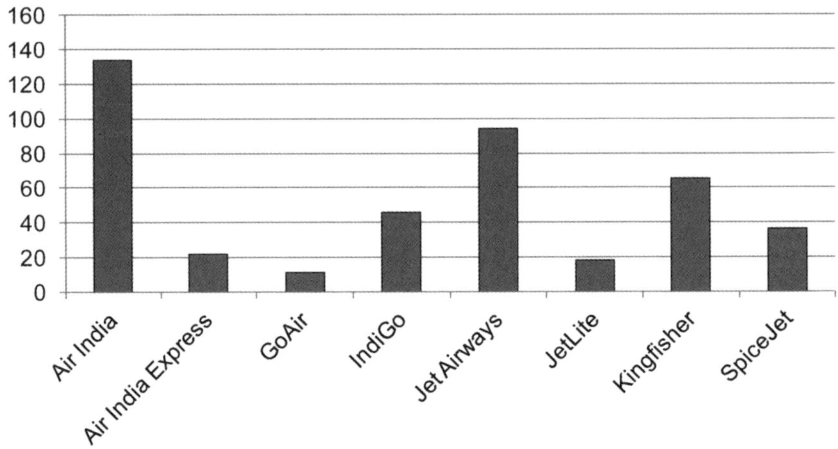

Figure 8.2 Indian scheduled airline fleets (December 2011)
Source: CAPA India and SITA 2012.

from booking to on-board the flight and many services like baggage allowance and airport check-in, free of charge.

IndiGo, SpiceJet and GoAir operate a combined fleet of new aircraft with an average age of one to three years (see Figure 8.2). There are 180 seats in most of these aircraft, sometimes more, allowing for maximum revenue per inch of aircraft space.

Despite 125 airports nationwide, India does not have the concept of multiple airports in one city. LCCs therefore do not have the option of operating into cheaper secondary airports, which means largely that the airport cost structure for them is the same as that of the FSC. The cost of operating to big Indian airports like Delhi and Mumbai is higher as they are private airports which are spending billions of dollars in modernization and consequently passing on the cost to the airlines and the passengers. India does not have the concept of airports being run by counties. The time it takes to turn around an aircraft after it lands and before it takes off for another flight is high in the country because of various infrastructure issues at the airports. With the advent of the LCCs in 2005, airports such as Delhi and Mumbai were caught unprepared. Flights had to remain in a holding pattern for up to 45 minutes at these airports before they could land, due to airspace congestion. Consequently, an LCC model in which maximum daily aircraft utilization is important to remain profitable, takes a hit. Well travelled international passengers, however, do not mind Indian discount carriers as they do not charge for oversized luggage, unaccompanied minors, or connecting to international flights, and offer their passengers two pieces luggage instead of the traditional 20 kg allowance. They are also more hospitable and the service is not treated just as a mode of transport. IndiGo, SpiceJet and GoAir also attempt to attract business travellers.

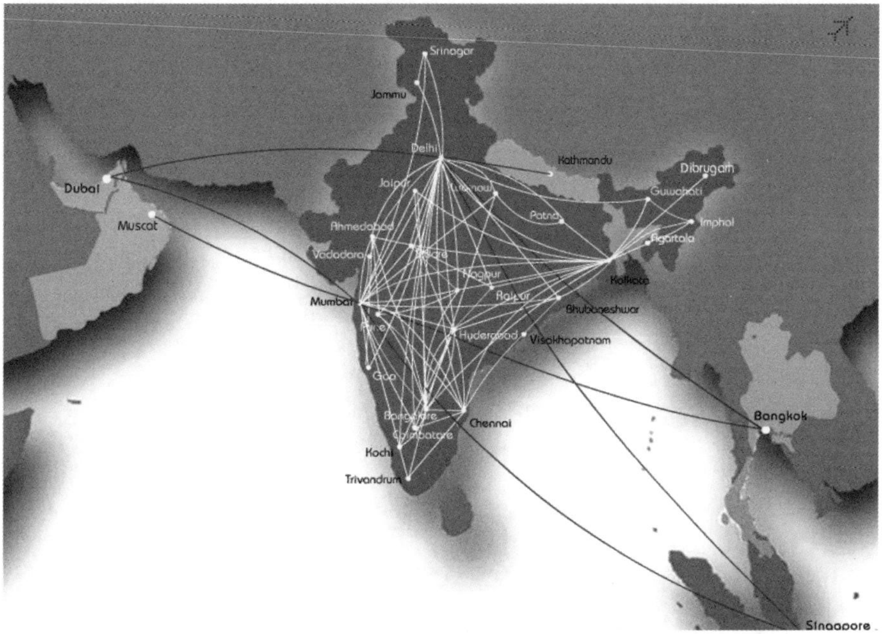

Figure 8.3 Route map of India's largest budget carrier IndiGo
Source: IndiGo 2012.

IndiGo and SpiceJet, for example reserve the first few rows for corporate travellers. GoAir on the other hand has a concept where for about US$20 the middle seat is left vacant.

Service levels amongst Indian LCCs and FSCs are also marginally different. For example, while all FSCs provide hot food and beverages onboard, LCCs like GoAir and IndiGo do not. In the last two years, SpiceJet has started serving hot food to cater for a growing demand and distinguish itself from other LCCs. However all three LCCs do not offer loyalty programme, in-flight entertainment systems, or lounge access.

In 2011, India's IndiGo topped the list of the fastest growing LCCs in the world with a 45.7 per cent surge in ASKs (AvailableSeatKilometres), taking it into the global LCC top 20 for the first time (CAPA 2011). SpiceJet is the third fastest growing LCC, while Mexico's Volaris splits the Indian carriers at second rank. IndiGo, now the country's largest LCC by passenger numbers, is becoming the airline to watch out for in the Asia Pacific region and, indeed, the world. The privately owned carrier, which shot into fame with a surprise order for 100 Airbus A320 aircraft at the Paris Air Show in 2005 at a list price of US$6 billion, has modelled itself in some ways on the business model of US-based LCC JetBlue. IndiGo was the last to start services of all the Indian airline launches in the last decade. 'We were the last to start up. By that time the slots were gone, the engineers were gone, the pilots were gone, the bays were gone – everything was gone', says Aditya Ghosh, president of IndiGo Consumer perception was loaded against LCCs, which were thought to be unreliable, operating shabby planes and known for chaotic check-in procedures (Shukla 2009).

However, IndiGo took some years to tweak the model, putting systems and processes in place under a team led by former US Airways executive Bruce Ashby, ex-Chief Operating Officer of North American Airlines Steve Harfst, and former head of aircraft financing at Emirates Riyaz Peermohammed.

Almost nothing is as important at the carrier as the focus on saving fuel, its biggest expense. While normally it is the pilot who decides what quantity of fuel to uplift for the operating flight, IndiGo measures the intake. For example, for a Mumbai–Delhi flight the airline takes an average of the fuel spent for the previous three months, considering weather conditions and airport/air space congestion. IndiGo then dictates exactly how much fuel is to be taken on for this sector.

Most of the operations are helped by the use of the software Aircraft Communications Addressing and Reporting System (ACARS), stationed at the airline's Global Business Park office in the outskirts of Delhi, which performs a minute-by-minute tracking of the aircraft that gives significant operational benefits. IndiGo gives a monthly roster to pilots well in advance, which keeps the crews happy as they can better plan for other work commitments and social life.

If there is a delay, for example due to fog, ATC, runway closures, VIP movements, or a technical glitch in the morning, it could affect the entire chain of flights throughout the day. Subsequently, crew members, ground staff and the pilots try their best to shave off time when possible. IndiGo also prefers the use of ramp positions instead of boarding stairs to allow more people to deplane and enplane the aircraft in a shorter time. In general boarding by ramp is faster than by stairs as old people and children may struggle to walk at speed.

While there may be limited opportunities to cut costs in the Indian operating environment, Indigo purchased 100 aircraft at a considerable discount, will sell the aircraft at the time of delivery, and lease them back from the same leasing company. The profit that is generated from this premium is used to subsidize their operations.

Conclusion

LCCs will continue to dominate the ongoing growth in the Indian skies. Yet their trajectory will be defined by the events of 2012. CAPA anticipates that it will be as challenging a year as 2004, when growth was about to peak. Indian aviation entered 2012 facing its most critical challenges since the advent of the 2004 industry reforms. Indian carriers are expected to lose US$2.5 billion in 2011–2012 alone on a total revenue of just under US$10 billion; a worse result even than in the financial year 2008/09, when traffic was declining and fuel prices spiked at US$150 per barrel.

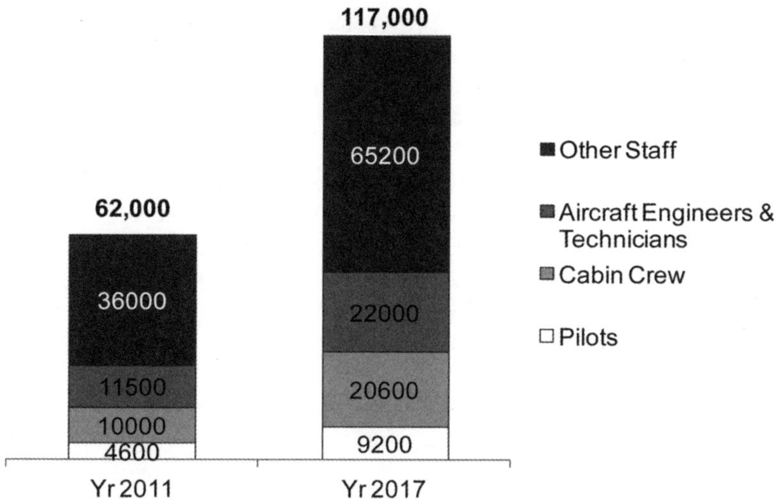

Figure 8.4 Workforce requirement for domestic carriers
Source: CAPA 2011b.

LCCs will need to balance their growth. Besides the shortage of workers and professionals, safety concerns, declining income and the lack of accompanying capacity, infrastructure, stiff competition and rising fuel costs will be challenging in the foreseeable future. A potential clearance by the Indian government to allow foreign airlines' investment in domestic carriers could herald a new chapter.

The opportunities for people with the right skills and experience exists and they are likely to expand further as Figure 8.4 shows. The workforce requirements of the airlines are expected to double over the next six years, but quality institutes to produce such manpower are in short supply in the country. Despite the challenges India's LCCs may face as they beat frequencies, the 2003–2007 liberalization of the Indian skies has ensured that growth will be unstoppable and irreversible for many more years.

References

CAPA (2011a). India's domestic aviation market shows rapid growth in first half, 14 September 2011. Sydney, Australia: Centre for Asia Pacific Aviation. Available at: http://www.centreforaviation.com/analysis/indias-domestic-aviation-market-shows-rapid-growth-in-first-half-58687 [accessed: 6 June 2012].

CAPA (2011b). *Investing in Human Capital*.Presentation at 6th ACI Asia Pacific Regional Assembly, 5–8 April 2011, New Delhi, India: Centre for Asia Pacific Aviation.

CAPA India and SITA (2012). *CAPA-SITA Report on Innovation & Technology: Opportunities for Transformation in Indian Aviation. March 2012*. Available at: http://www.centreforaviation.com/files/analysis/69963/CAPA-SITA%20 Report%20on%20Innovation%20&%20Technology.pdf [accessed: 12 May 2012].

Civil Aviation Ministry (2011).*Report of Working Group on Civil Aviation for formulation of Twelfth Five Year Plan (2012-17)*. Available at: http:// civilaviation.gov.in/cs/groups/public/documents/newsandupdates/moca _001377.pdf [accessed: 12 April 2011].

DGCA – Directorate General of Civil Aviation (2012). Market share of scheduled domestic airlines. Available at: http://dgca.nic.in/reports/pass-ind.htm [accessed: 6 June 2012].

Economic Times (2011a). *India's GDP growth rate to slip below 7%: Moody's*, 21 December 2011. Available at: http://articles.economictimes.indiatimes. com/2011-12-21/news/30542657_1_bond-ratings-growth-rate-stable-outlook [accessed: 6 June 2012].

Economic Times (2011b). Internet base in India crosses 100 million mark: What does it mean for businesses that can ride on this? 15 December 2011. Available at: http://articles.economictimes.indiatimes.com/2011-12-15/news/ 30520358_1_e-commerce-indian-internet-space-internet-and-mobile-association [accessed: 6 June 2012].

Economist online, *The* (2012). *Daily Chart: Top Flights: Where are the World's Busiest Airline Routes?* 14 May. Available at: http://www.economist.com/ comment/1416808 [accessed: 7 June 2012].

Gopinath, G.R. (2010). Simply fly: A Deccan odyssey. New Delhi, India: HarperCollins.

IndiGo (2012). Network map of the airline. Available at: www.goindigo.in [accessed: 12 January 2012].

Hindu, The (2011). India poised to be among top 5 aviation nations in next 10 years: *Pratibha Patil*, 17 October 2011. Available at: www.thehindubusinessline. com/industry-and-economy/logistics/article2545646.ece?ref=wl_banking_art [accessed: 12 February 2012].

Pilling, D., Hille, K. and Kazmin, A. (2011).Asia: The rise of the middle class, hope for a future of self-generated growth. *Financial Times*, 4 January 2011. Available at: http://www.ft.com/intl/cms/s/0/5841236e-183a-11e0-88c9-00144 feab49a.html#axzz1qnQax22T [accessed: 12 February 2012].

Shukla, T. (2009). IndiGo profits from cost cuts, in *livemint.com – The Wallstreet Journal*, 19 October 2009. Available at: http://www.livemint. com/2009/10/19004355/IndiGo-profits-from-cost-cuts.html [accessed: 12 February 2012].

Shukla, T. (2011). Praful Patel blamed for grounding Air India, *livemint.com – The Wallstreet Journal*, 9 September 2011. Available at: http://www.livemint. com/2011/09/09001218/Praful-Patel-blamed-for-ground.html [accessed: 12 February 2012].

Sinha, S. (2012). Kingfisher Airlines safety an issue: DGCA, *The Times of India*, 5 January 2012. Available at: http://articles.timesofindia.indiatimes.com/2012-01-05/india-business/30592650_1_kingfisher-airlines-dgca-pilots [accessed: 12 February 2012].

Webb, M. (2011). Asian low-cost carriers: Identifying winners and losers, *HSBC Global Research*. Available at: http://www.research.hsbc.com/midas/Res/RDV?p=pdf&key=g3V7aPecWv&n=310173.PDF [accessed: 25 October 2011].

Chapter 9
Low Cost Carriers in Australia and New Zealand

Michael Lück and Sven Gross

Introduction

In the past few years throughout North America and Europe, carriers launched as low cost carriers (LCCs) were mostly able to establish themselves quickly in the market due to the inflated cost structures of their traditional competitors. In addition, changes in consumer preferences favoured these LCCs. Generally, consumers had fared well with discount products in other markets (e.g. last-minute travel, electronic products, hotels). However, in North America and Europe (for the most part) incumbent carriers have been able to come to grips with attacks from a host of start-up airlines, successfully resorting to cost-cutting schemes that would bring about a reduction of much of the initial differences in costs and, in turn, prices. Today, consumers opting for traditional airlines are generally able to purchase cheap tickets, usually by taking advantage of offerings of unsold seats available from the airlines' homepages.

While in the USA and Europe low cost airlines (LCAs) were already established in the 1970s and 1990s, respectively, the low cost industry in Australia and New Zealand has not been widely developed until the beginning of this century, but has shown growth since. Today the low cost market accounts for 24.9 per cent of the total seats within the Asia Pacific region, compared to 35.3 per cent within Europe and 30 per cent within North America, which is a rapid growth in Asia/Pacific, from 1.1 per cent to 24.9 per cent in just over one decade.

The Aviation Profile in Australia and New Zealand

Significance of Tourism and Air Travel

Market liberalization policies have enabled continued economic growth in Australia and New Zealand in the past two decades. Travel and tourism have become key economic drivers for the Australian and the New Zealand economies. In Australia, tourism directly employed 4.7 per cent of the workforce, contributed 2.5 per cent to gross domestic product (GDP) and was worth around 8 per cent of

Table 9.1 Travel times from Sydney to other major cities in Australia

	Air	Rail	Coach
Canberra	0.45	4.00	5.00
Adelaide	1.40	25.00	22.00
Brisbane	1.20	15.00	15.00
Darwin	5.00	—	92.50
Melbourne	1.10	10.00	14.00
Perth	4.00	65.00	56.00
Hobart	2.05	—	—

Source: Allo' Expat 2009.

exports in 2010–2011. In New Zealand, tourism directly or indirectly employed 9.4 per cent of the workforce, contributed 8.6 per cent to GDP and 16.8 per cent to New Zealand exports in the same period (Statistics New Zealand 2011, Australian Bureau of Statistics 2011).

The region's population is spread over a relatively large area of land and sea, which is why people rely heavily on aviation to get from place to place. Especially in Australia, with many regions being remote and desert terrain, flying is the most convenient transport option for reaching many parts of the country. Table 9.1 substantiates this fact by comparing the approximate travel times from Sydney to other major cities by the different means of transport. In addition, LCCs do not have to compete with, for example, a high speed rail network.

New Zealand has been Australia's number one source market since 1999. In the year ending September 2011, Australia received 1,164,066 visitors from New Zealand, which was the largest source of visitors, followed by the United Kingdom (631,256), the USA (456,005), and Japan (342,740) (Tourism Australia 2011). Australia is New Zealand's largest and steadily increasing inbound tourism market. In 2011, almost 45 per cent of all arrivals originated in Australia, amounting to a total of 1,157,962 visitors, which is an increase of 3.2 per cent from the previous year. In the future visitor arrivals are expected to rise on average by 4 per cent annually (from 2008 to 2014) (Tourism New Zealand 2008; Statistics New Zealand 2012).

The tourism trends observed in Australia and New Zealand are broadly similar. The trans-Tasman aviation route (between New Zealand and Australia) is Australia's and New Zealand's busiest in terms of capacity and frequencies (see Table 9.2). Seat capacity peaked in 2005, following the entrance of Emirates and Pacific Blue in 2004, and Jetstar in 2005. Air New Zealand remains the dominant airline on the trans-Tasman route (46 per cent). Qantas operated around 17 per cent of all flights in 2009 (including services operated by JetKonnect, a subsidiary of Qantas), Pacific Blue 15 per cent, and Jetstar 8 per cent.

Table 9.2 Trans-Tasman Aviation market share 2007

Airline	2003	2004	2005	2006	2007	2008	2009
Air New Zealand	37%	36%	40%	42%	42%	47%	46%
Qantas	33%	28%	25%	21%	19%	19%	17%
Emirates	3%	7%	8%	9%	11%	11%	9%
Freedom Air	18%	15%	13%	10%	9%	—	—
Pacific Blue	0%	6%	7%	7%	8%	11%	15%
Jetstar	0%	0%	0%	6%	6%	6%	8%
Other	9%	8%	7%	5%	5%	6%	5%
Total	100%	100%	100%	100%	100%	100%	100%

Source: Tourism Australia 2008, 2010.

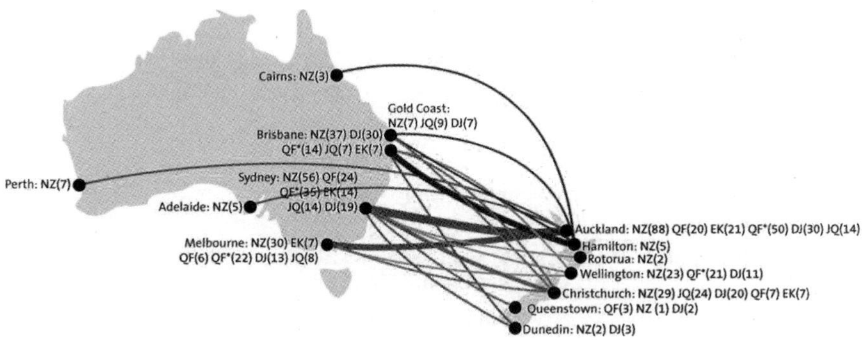

Figure 9.1 Direct services across the Tasman Sea
Source: Tourism Australia 2010.

> Qantas and Freedom Air both lost market share to the relatively new entrants Pacific Blue, Jetstar and Emirates. In 2007, direct capacity on the New Zealand–Australia route fell 6 per cent year-on-year, while load factors increased 8 percentage points. Capacity reductions were led by the Air NZ group (with Freedom Air down 18 per cent) in early 2007 and the suspension of Qantas's Auckland–Adelaide services in July 2007. Jetstar and Pacific Blue were the only two carriers to increase capacity on this route in 2007 (Tourism Australia 2008).

In March 2008, Freedom Air withdrew services and was integrated back into mother company Air New Zealand, which replaced most of these services. Low cost airlines Jetstar and Pacific Blue increased capacity across the Tasman significantly (32 per cent and 59 per cent, respectively), resulting in an overall increase of 4 per cent year-on-year in 2009 (Tourism Australia 2010). Figure 9.1 details direct services across the Tasman Sea.

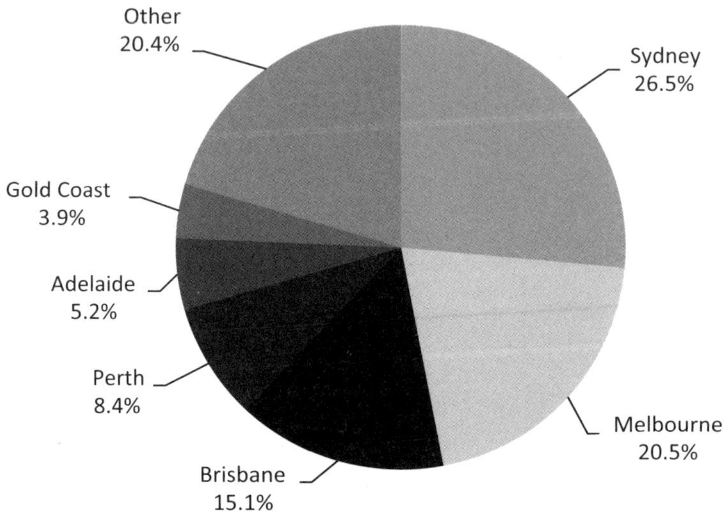

Figure 9.2 Airport shares in air passenger movements 2009–2010
Source: BITRE 2011.

Airports in Australia and New Zealand

Australia has a good complement of airports, both international and domestic, as well as many smaller, regional terminals (101 airports in total). With each state capital including at least one major international airport, all important urban centres are easily accessible by air. Sydney International Airport is Australia's premier airport (see Figure 9.2), with almost 35.5 million passengers arriving and departing from this airport every year (BITRE 2011).

New Zealand offers a total of 36 international and domestic airports. Auckland Airport, New Zealand's largest and busiest airport based on total passenger numbers and cargo, serves approximately 70 per cent of the international arrivals and departures. Airports in Christchurch, Wellington, Dunedin, Queenstown, Hamilton and Rotorua also receive flights from other countries, mostly Australia. Christchurch is New Zealand's second largest airport, and Wellington International Airport is a major domestic hub in the regional and national transport system, as well as providing international services to Sydney, Melbourne and Brisbane (Tourism New Zealand 2009).

The Emergence of Low Cost Carriers in Australia and New Zealand

Over the past four decades the passenger aviation industry has been subjected to several waves of innovations. Three very important innovations are the political liberation in the 1970s and 1980s, the rise of LCCs, and the widespread application of the Internet.

> Perhaps no international air travel market has been more radically affected by these innovations than the set of trans-Tasman routes linking the major cities of Australia and New Zealand. Few aviation markets are now so open to competition, including competition from airlines based in third countries. And the relatively short (three to four hours) and simple (point-to-point, with, of course, no stopovers en route) nature of the product has facilitated the entry of LCC services and encouraged the adoption by consumers of 'do it yourself' comparison shopping on the internet (Hazledine 2008: 335).

The deregulation of domestic aviation in New Zealand commenced in 1983 and was completed in 1990 with the abolition of air services licensing. In 1986, New Zealand already allowed up to 100 per cent foreign ownership of domestic airlines. Australia initiated the deregulation process in 1990 and successfully completed it in 1995. Prior to this, Qantas and Ansett dominated the market (Forsyth 2003). The deregulation of Australia's domestic airlines meant that flight services were getting more competitively priced (BITRE 2006, Francis et al. 2006, Statistics New Zealand 2009).

The majority of LCC development in the region has taken place on the trans-Tasman route, as well as domestically. Thus, this chapter will focus on these areas.

Developments in past years are characterized by airlines entering and withdrawing from the various international markets (see Table 9.3). Some of the airlines were founded as low cost subsidiaries of traditional airlines (e.g. Jetstar by Qantas and Freedom Air by Air New Zealand) in order to react to offers of emerging LCCs (see Figure 9.3).

Important recent developments in Australia are the new entrants Virgin Blue, Jetstar, and Tiger Airways. Virgin Blue was founded in August 2000 and has grown into Australia's second-largest airline (by fleet size). Jetstar commenced flying in May 2004 on some trunk routes, and on some routes from secondary airports (BITRE 2006, Forsyth 2007). Qantas established Jetstar as a tool to antagonize the rise in the domestic Australian market of Virgin Blue (Qantas Airways 2007). Tiger Airways, a subsidiary of Tiger Aviation (partly owned by Singapore Airlines), entered the market in November 2007 with services in the Australian domestic airline market. The airline is based in Melbourne, with its main hub at Melbourne Airport (BITRE 2008). Tiger Airways Australia is connected to the parent company's extensive Asian network by flights from Perth to the main hub Singapore, serviced by Tiger Airways Singapore.

Pacific Blue was founded in 2003 and commenced operations mainly out of Christchurch. Pacific Blue is a subsidiary of Virgin Blue, and was launched in order to position itself as a low fare competitor to Air New Zealand and Qantas on trans-Tasman routes. In August 2007, Pacific Blue initiated services within New Zealand, but withdrew from the New Zealand domestic market in 2010 (New Zealand Herald 2010). For flights to the South Pacific, Pacific Blue was formed as a subsidiary, and in 2009 a long haul arm was launched under the name V Australia. After a long term dispute over the use of the Virgin brand in Australia

Table 9.3 Developments in the Australia/New Zealand low cost market (1990–2009)

Name	Country of origin (base)	Year of foundation	Start of flight operations	Withdrawal from the market	Parent/holding company
Compass Airlines	Australia	1990	1990	1991	
Compass Airlines II	Australia	1992	1992	1993	
Impulse Airlines	Australia	2000	2000	May 2004	(acquired by Qantas 2001)
Virgin Australia	Australia	2000	2000	—	Virgin Blue Holdings Limited
Jetstar Airways	Australia	2003	May 2004	—	Qantas Airways
Tiger Airways Australia	Australia	March 2007	November 2007	—	Tiger Aviation
Kiwi Travel International Airlines	New Zealand	1994	1994	September 1996	
Freedom Air	New Zealand	1995	1995	March 2008	Air New Zealand
CityJet	New Zealand	1999	May 1999	November 1999	
K2000	New Zealand	2000	—	2000	Central Pacific Airlines
Pacific Blue	New Zealand	2003	2004	—	Virgin Blue
AirAsia X	Malaysia	2007	2007	—	Franchise of AirAsia

Source: Based on airline websites (status: February 2009).

Figure 9.3 Air New Zealand founded its low cost subsidiary, Freedom Air, in response to emerging low cost airlines

Source: Michael Lück.

was settled, Virgin announced that all its regional operations (Virgin Blue, Pacific Blue, V Australia) were to be rebranded and marketed under the new name 'Virgin Australia' from 2011 onwards.

More recently, V Australia, Jetstar and AirAsiaX commenced international long haul, low cost services from Australia and New Zealand. However, these services have become very volatile due to high fuel prices and other external factors such as the 2010 and 2011 earthquakes in Christchurch. Subsequently, V Australia repositioned itself, discontinued flights to Thailand and South Africa, instead entered into an extensive agreement with Etihad Airways, and began operating flights to Abu Dhabi. After less than 18 months, AirAsiaX ceased operations at Christchurch, following the weakened demand due to the earthquakes. However, AirAsiaX continues services to the Australian ports of Perth, Darwin, Melbourne and Gold Coast. Jetstar is the only remaining LCC in the region that offers a comprehensive long haul network, including various points in Asia and Hawaii.

Components of the Business Model

A close look at the airline industry shows that there is a wide range of different implementations of the low cost strategy, and that the boundaries between traditional carriers and LCCs are blurred. The many configuration opportunities of the single business components will be analysed further for the airlines which operate within Australia and New Zealand, and/or offer trans-Tasman services. Subjects of procurement and suppliers, as well as process management and marketing, and opportunities for their practical implementation will be pointed out.

The central element of a low cost strategy is a continuous analysis and control of the most important costs within a company and an efficient asset management. Therefore the product, distribution and communication policies, in addition to the pricing policy, need to be considered.

Procurement

Aircraft Aircraft can be procured through purchase or leasing of both newly manufactured and second-hand aeroplanes. To finance their aircraft and required technical equipment, LCCs mainly use the leasing option. An important factor for the success of the low cost segment in Europe was the opportunity to buy second-hand aircraft. In the aftermath of the worldwide aviation crisis following 11 September 2001, many used aeroplanes were available on the international market at very low prices, which greatly benefitted LCCs (Bjelicic 2004). Most LCCs in Australia and New Zealand started operations with leased and/or owned second-hand aircraft. In some cases they commenced operations with aircraft of their respective parent airline's fleet (e.g. Jetstar, Freedom Air, Pacific Blue). However, these airlines increasingly purchase brand new aircraft from the manufacturers, supported by good deals due to high order numbers. In some cases, the airlines sell their new aircraft to a leasing company and lease them back. The trend to the purchase or lease-back of new aircraft results in modern fleets, which has positive effects on the operating costs (e.g. fuel efficiency, reliability, maintenance cost).

More recently, larger orders have been placed by LCCs in Australia. For example, Qantas has confirmed that a total of 53 aircraft have been earmarked for its low cost subsidiary Jetstar (Business Day 2009). A uniform fleet of the same aircraft type or family leads to cost savings for personnel training and more flexible operational planning, as flight and technical staff are subject to identical qualification standards. This also allows cost savings in the field of maintenance and servicing, for example with respect to spare part stock management. However, disadvantages may arise due to dependence on a sole manufacturer (= supplier power). Also, flight scheduling advantages for LCCs result from a uniform fleet, since the different performance features and capacities of various aircraft types need not be taken into account. However, any possible variation in demand is impossible or difficult to compensate due to rigid capacities. The use of uniform aircraft in Australia and New Zealand, such as Boeing B737-800 and Airbus A320, is noticeable with LCCs currently operating in this market (see Table 9.4).

Airports Airports are an important element within the business model of European LCCs, since they provide the necessary infrastructure. Savings on

Table 9.4 Fleet structure of Australian low cost carriers

Name	No. of aircraft	No. and type of aircraft	Different types	Orders
Virgin Australia	82	11 B737-700 46 B737-800 18 Embraer 190 2 A330-200 5 B777-300	4 main types (2 sub-types of B737)	1 Boeing 737-800 1 Boeing 737-700 1 Airbus 330-200
Jetstar Airways (incl. Jetstar Asia and Jetstar Pacific)	87	65 A320-200 6 A321-200 11 A330-200 5 B737-400	3 main types (2 sub-types of A320 family)	12 A320-200 9 A321-200 2 A330-200 15 B787-800
Tiger Airways Australia (incl. Tiger Airways, Singapore and Indonesia)	33	33 A320-200	1	1 Airbus 320-200

Source: Based on airline websites and Airfleets (2012).

fees are often directly negotiated between the airlines and the respective airport, and may include various quantity discounts or marketing grants. Although the negotiated fees do not allow all airports to cover their costs, they sometimes agree to the conditions dictated by a LCC because of the beneficial direct and indirect economic effects for the airport and the surrounding region. Experiences from LCCs in Australia/New Zealand are not known to the authors, but it is less likely that major airports will agree on concessions, given the increased demand by the LCCs and the lack of secondary airports in this region.

Outsourcing, catering, waste disposal, fuel Since most LCCs do not have the required economies of scale, technical services such as maintenance and repair are outsourced to specialists who are able to carry out these jobs with greater cost-effectiveness, thus avoiding extra fixed costs for the airline company (Pompl 2007, Doganis 2001). Also, passenger handling and other ground services ('ground handling') are assigned to external handling agents. This allows LCCs to drastically reduce station facilities and staff. It is therefore typical for LCCs to incur low costs for ground staff and office/waiting rooms (Doganis 2001). Some airlines have gone a step further by introducing automatic and/or Internet check-in procedures (e.g. Jetstar, Tiger Airways) or check-in via mobile phone (e.g. Virgin Australia). Such measures would also be relatively easy to implement for all LCCs in Australia and New Zealand. Customers of the LCCs in this region are able also to book, manage and cancel their bookings online. Jetstar's ground-handling services are provided by its parent Qantas, and Tiger Airways contracts Singapore Airlines with maintenance services (Qantas 2008, Tiger Airways 2009).

Although most LCCs do not include catering in the fare, a selection of food and drinks is often offered for purchase, either at the time of booking or onboard.

This not only saves catering costs but it also generates additional revenue. Another positive effect is generated by reduced turnaround times of aircraft since less time is needed for loading. For example, Jetstar has reduced its turnaround times for domestic New Zealand flights by an additional five minutes, which allows for an additional rotation per day. In addition, catering is not complimentary on most Australian/New Zealand LCCs, with the exception of Jetstar and Virgin Australia, which include food and beverages in their StarClass and Business, respectively (see Table 9.6). Cost savings are also achieved by LCCs with respect to waste collection and cleaning inside the cabin, since these services are partly undertaken by cabin staff, thus saving costs for external ground-handling contractors.

Process Management

Strategic flight scheduling The established airlines focus on network-based structures, while European LCCs concentrate on point-to-point operations between passenger-intensive economic centres. Airports are mainly offered in parallel markets, i.e. cities with several airports which are not yet being served by competitors, or secondary airports near larger economic centers (Pompl 2007). Because they use mainly secondary airports with no capacity restrictions, and without congestion, LCCs usually do not have problems in obtaining slots. Table 9.5 provides an overview of the LCCs in Australia/New Zealand, the network configuration and their destinations and routes.

While Tiger Airways offers point-to-point services only, Virgin Australia and Jetstar also offer connecting flights (the latter also for parent Qantas). Virgin Australia provides connections to Hawaii, Abu Dhabi, Phuket, the South Pacific Islands, New Zealand, and within Australia. Virgin Australia and Hawaiian Airlines also signed an interline agreement, which 'will enable travel agents to issue a single ticket for travel on both airlines, from anywhere in the Virgin Blue

Table 9.5 Destinations, routes and network configurations

Name	Destinations 2008	Network configuration
Virgin Australia	55 (39 Australia, 6 New Zealand, Vanuatu, Cook Islands, Solomon Islands, Fiji, Tonga, Samoa, Papua New Guinea, Indonesia, Thailand, USA, Abu Dhabi)	point-to-point, connecting flights, feeder
Jetstar Airways	55 (Australia, New Zealand, Asia Pacific)	point-to-point, feeder (also for parent Qantas)
Tiger Airways Australia	35 (6 Australia, 29 Asia)	point-to-point, connecting flights

Source: Based on airline websites (status: March 2012).

Figure 9.4 Virgin Australia serves primary airports and competes directly with Qantas

Source: Michael Lück.

network to anywhere in the Hawaiian Airlines network. It will also enable Hawaii-bound passengers around Australia to check their luggage all the way to Honolulu when they check in for their Virgin Blue connecting flight' (Virgin Blue 2006). Virgin Australia has also signed a comprehensive partnership with Etihad Airways of Abu Dhabi and entered an alliance with Air New Zealand (CAPA 2010).

New Zealand and Australia do not have many secondary airports, which makes competition with the traditional full service airlines (FSAs) more difficult, as LCCs have to utilize the same primary airports. In several cases, Jetstar has chosen 'to avoid head to head competition on routes with Virgin Blue and its parents, Qantas, by flying from a secondary airport' (Forsyth 2007: 92). In contrast, Virgin Australia serves primary airports and competes directly with Qantas (see Figure 9.4).

LCCs in Australia are challenged with different environmental conditions, compared to LCCs in Europe, Asia and North America. Due to the size of the country (Australia) and the geographical remoteness, routes tend to be much longer compared to the routes of European LCCs. The shortest flights between Australia and New Zealand, or either of the countries and the Pacific Islands take about three hours. Thus short turnaround times and no frills (some frills must be maintained on longer flights), important determinants for cost benefits, tend to become more difficult to manage. Other exceptions are the recently emerging long haul operations to Europe, Asia and the USA (e.g. V Australia (now Virgin Australia), Jetstar and AirAsiaX).

Personnel policy Due to their concentration on core competencies and related outsourcing operations, LCCs managed to downsize their workforce to a minimum. Owing to limited onboard services, fewer in-flight cabin staff are required. Since most of the LCCs have been on the market for a relatively short period of time, and given the fact that they hired staff during economically difficult times and unions rarely had a say, it was possible to achieve low pay agreements with longer working hours. Other factors which contribute to cutting costs are the maximum utilization of permitted working hours within the legal limits, avoidance of voluntary social charges (such as holiday allowance or Christmas bonus), as well as flight operation with minimum staff (Bjelicic 2004, Goettert and Schmidt 2005).

Marketing

Product policy The traditional distinction between different booking categories (booking classes) of legacy carriers is not applied by most LCCs. Australian/New Zealand LCCs generally offer two classes (one standard economy and one premium class), with the exception of Tiger Airways (see Table 9.6). Virgin Australia is now promoting a more differentiated 'New World Carrier' strategy, a business plan which is aimed at the contemporary business traveller. In this regard Virgin Australia recently launched Australia's first Premium Economy product (CAPA 2008).

Seat density, which is a product feature, is determined according to principles designed to optimize revenue. Tighter seating (typical seat pitch is 29–30 inches instead of the usual 31–34 inches of traditional airlines) leads to a restriction of product quality, while allowing an increase in transportation capacity. Consequently, costs per seat and flight prices drop and the priority service feature desired by customers is achieved (Sterzenbach and Conrady 2003). Virgin Australia and Jetstar offer a seat pitch of at least 30–31 inches on their newer aircraft, which is a comparatively large seat pitch for an LCC (see Table 9.6). Australia offer a so-called 'Blue Zone', where for an additional charge (AU$25–45), passengers can book in advance an extra roomy over-wing exit row seat on Embraer E190 and Boeing 737 aircraft. The Blue Zone seating offers the equivalent of a 39-inch seat pitch. In addition, Virgin Australia introduced a business class on domestic and international routes. Depending on aircraft type and configuration, business travellers are either seated in a separate business class, or have economy seating with a 'middle-seat-empty' guarantee.

Seat density describes the total actual number of seats of an aircraft, compared with the total number of seats possible (as per manufacturer specifications). The total maximum capacity at which the airline operates the aircraft is calculated as a percentage. Table 9.7 shows that the airlines operating with 100 per cent capacity are Jetstar and Tiger Airways. A study by J. Edwards (2008: 7) argues that 'FSCs [full service carriers] operate their aircraft at as much as 33 per cent less than the maximum capacity levels employed by the highest-ranking LCC', which significantly adds to the operating cost per seat.

Table 9.6 Service classes, price discrimination and range of services

	Virgin Australia	Jetstar Airways	Tiger Airways
Service Classes (number and name)	3 Business Premium Economy Standard Economy	2 Economy StarClass	1 Economy
Price discrimination (fare name)	Saver Flexi Premium Economy Business	Starter Plus Max Business Business Max	Internet Discounted Fare Regular Fare
Seat Pitch	Dedicated business class seats Premium Economy 33+ inches (84 cm) Standard Economy 31–33 inches (78–84 cm)	StarClass 38 inch (96,5 cm) Economy class 30–31 inches (76–78 cm)	28,5 inches (72.5 cm)
FFP	Velocity Rewards (with partner National Australia Bank offer of a companion credit card).	Qantas FFP for Plus, Max and Business Max fares	Not available
Inflight entertainment	Monthly in-flight magazine, Live2air in-flight entertainment (24 live channels; $4.95 short flights, $9.90 longer flights; advertising channel and flight track channel free of charge), on international flights digEplayer (blockbuster movies, TV shows, music for a hire fee of $15). Entertainment included in Premium Economy and Business class.	Business class: Video on Demand Unit and headset (movies, TV programs, music videos). Economy class: limited quantity of Video on Demand units available for hire; several screens along the aisles which will show films and TV programs; passengers can purchase reusable headsets; audio programs.	Quarterly in-flight magazine
Range of services (e.g. catering)	Economy class: not complimentary. Premium Economy and business class: complimentary.	Economy class: not complimentary, depending on fare. Business class: complimentary.	Not complimentary

Source: Based on airline websites (status: March 2012).

Table 9.7 Seating density (selected aircraft)

Name	Aircraft types	Total number of this aircraft in fleet	Number of seats on aircraft	Maximum seat capacity for aircraft type	Percent of max. seat capacity
Virgin Australia	B737-700	11	140	149	94
	B737-800	46	176	189	93
Jetstar Airways	A320-200	65	177/180*	180	98–100
	A321-200	6	210/214	220	95–97
Tiger Airways Australia	A320-200	33	180	180	100

Note: * 177 passengers within Australia and New Zealand and 180 passengers on intra-Asia flights.
Source: Based on airline websites (status: March 2012).

Frequent flyer programmes are part of the standard services offered by traditional airlines. Since cost reduction is given priority by LCCs and those programmes involve high administration costs, 'classic' LCCs have not introduced such programmes. However, LCCs in Australia and New Zealand operating within the domestic market (with the exception of Tiger Airways) offer an own frequent flyer programme, or participate in an external scheme (see Table 9.6).

Pricing policy Since the feature of low fares is by far the most important feature for consumers, pricing represents the most powerful marketing instrument for LCCs, while being a long term factor for customer loyalty. For pricing, LCCs use a mixed calculation, i.e. the average ticket price determined by cost accounting procedures (fixed and variable costs for a certain occupancy rate) serves as a basis for the offered prices. Flights are sold at different prices, with some of the tickets being distributed at a loss, which in turn is compensated for by the high-priced tickets. In contrast to traditional pricing policy, a time-related price discrimination is implemented through penetration pricing.

At first, low base prices (initial prices) are determined, corresponding to the strategic pricing policy aims, i.e. market-orientated pricing based on competition and demand is undertaken. These base prices are communicated to the consumers in order to encourage them to book early. In contrast to 'last-minute prices', which are perceived by customers as being reduced as time goes on, thus increasing the occupancy risk, this price system conveys to the customer the idea of a 'price guarantee', i.e. there will be no cheaper prices for a certain flight at a later point in time. As booking goes on, pricing becomes more cost-orientated and prices increase as the departure date draws closer. There may be a constant increase in prices, or prices may be adjusted according to the revenue management parameters (e.g. booking details from the past, prognostics or price flexibilities) and the actual booking situation.

Consequently, LCCs practise a dynamic and flexible pricing policy with a mixture of cost- and market-orientated elements. In a first step, a specific low price contingent is determined. Its size varies from one airline to another, comprising between 10 per cent and 70 per cent of the offered seats, averaging 20–30 per cent (Ramm 2002). When this contingent is sold, prices increase in steps and may reach, or even go beyond, the prices of traditional airlines.

Fares of LCCs in Australia and New Zealand appear to be set by LCCs to match competitors' fares, and often undercut these. Most LCCs in Australia and New Zealand also offer different fares with specific terms and conditions, as illustrated in Table 9.6.

Distribution policy LCCs sell their services through few distribution channels, mainly directly through the Internet or booking machines (direct self-distribution) or through their own centralized call centres established at cost-competitive locations. More savings of distribution costs are achieved by issuing tickets together with the boarding pass, and by using electronic tickets. Printing and material costs are even shifted onto the passengers (e.g. print-out of tickets/boarding passes). The collection of payment, which has traditionally been through intermediaries or special accounting agencies (e.g. Airplus), is practised cost-effectively through credit and debit cards, creating a positive impact on the LCCs' solvency situation (Pompl 2007). Virgin Australia and Tiger Airways, for example, have an effective distribution model: over 90 per cent (Virgin Australia) and 85 per cent (Tiger Airways) of tickets sold are via the Internet (InvestSMART 2009, Tiger Airways 2009). LCCs in Australia/New Zealand are bookable via GDS, and some via intermediaries (such as Expedia), as illustrated in Table 9.8.

Table 9.8 Methods of booking and check-in services

	Method of booking				Check-in			
	Internet	Call centre	GDS	Intermediaries	Online free seat selection	Online chargeable seat selection	Telephone	Self-service kiosk
Virgin Australia	✓	✓	✓	✓	✓	N/A	N/A	✓
Jetstar Airways	✓	✓	✓	✓ (under Qantas flight number)	✓	✓	N/A	N/A
Tiger Airways Australia	✓	✓	N/A	N/A	N/A	✓	N/A	N/A

Note: ✓ = Airline offers service (in at least one area of its operations); N/A = service is not available.
Source: Based on airline websites (status: March 2012).

Communication policy The communication policy is designed to inform (potential) customers about the available services offered by a LCC. Public relations work is partly carried out through spectacular campaigns, helping LCCs attract great public interest and gain free publicity. Sales promotion should be considered in close relationship with this. This includes enabling of both self- and third-party distribution channels, while it is also directed towards the end consumer. LCCs run various sales promotion campaigns such as the sale of tickets at a symbolic price (e.g. for 1 dollar) or giving away tickets free of charge. Furthermore, communication is practised through intensive advertising addressed to the end consumer. Advertising is focused on the price, in which the net price is mentioned most of the time. Any extra charges such as general taxes, handling or safety duties, and fuel taxes are charged separately and not mentioned at all or only as a footnote in advertisements. LCCs prefer advertising in daily newspapers ('daily prices need a daily press') and on the Internet, for example through social media. The communication concept incorporates both classical advertising through selective presence in relevant markets and the use of modern information technologies. But there are also innovative approaches which are being followed by low cost airlines. For example, a documentary series run by Tiger Airways ('Air Ways'), which is about the life of passengers and staff of the airline, reached an audience of up to 1.5 million viewers.

In addition to corporate behaviour and corporate communication, the corporate identity encompasses the external image (corporate design), including the branding of aircraft. LCCs use the fuselage to communicate booking phone numbers (Virgin Australia) or Internet domains (tigerairways.com).

Jetstar also pursues innovative approaches in its communications policy, with its own TV series ('Going Places'), its major sponsorship of the National Rugby League team (the Jetstar Gold Coast Titans), and the Australian TV programme 'The Morning Show'. The use of aeroplanes as advertising space for third companies was also put into action by Jetstar (pizza advertisement). In addition, Jetstar provided sponsorship to AFL matches held in Tasmania, was a corporate partner of the Geelong Football Club's football volunteer programme, and is a major supporter of the Tasmanian Symphony Orchestra and its school education programme (Qantas 2008).

Conclusion

The LCCs in Australia and New Zealand are another example of the worldwide trend towards airlines offering heavily discounted fares. The local airline market has initially been dominated by the national airlines Qantas and Air New Zealand for many years. However, after some unsuccessful initial attempts, the last decade has witnessed a major restructuring. The former duopoly of Qantas and Air New Zealand was stirred up by the emergence of LCCs, significantly lowering fares. These discounted airfares have generated additional competition, providing greater deals for consumers. This has made travel more attractive and will further accelerate the positive growth rates in the tourism industries in Australia and New Zealand. In a liberalized and more competitive environment, the new LCCs achieved significant market shares of the region's aviation industry in just a few years.

Having started as a low complexity and low cost airline, Virgin Australia evolved into a highly integrated air services provider. It benefits from a highly competitive cost structure, a modern and efficient fleet, flexible workplace arrangements and an effective and streamlined distribution model. It established a comprehensive domestic network with connectivity, great flight frequencies, a loyalty programme, lounges and other customer services, and already launched two international joint ventures. Tiger Airways remains the only 'real' no frills airline (e.g. one aircraft model, one booking class, no services included, no in-flight entertainment except an in-flight magazine).

The future development of the low cost business is expected to generate further demand on the trans-Tasman route, as well as increasing international outbound travel to other countries worldwide. Many LCCs have already altered airline strategies, offering long haul routes to Asia and North America. It is evident that the airline market in Australia and New Zealand has significant growth potential. However, the fierce current competition means it will be very difficult for new entrants into this industry.

Acknowledgement

An earlier version of this chapter was published in the *European Journal of Transport and Infrastructure Research* (2011).

References

Airfleets (2012). Airfleets. Available at: http://www.airfleets.net/home/ [accessed: 15 March 2012].

Allo' Expat (2009). *International Airport in Australia*. Available at: http://www.australia.alloexpat.com/australia_information/international_airport_australia.php [accessed: 5 February 2009].

Australian Bureau of Statistics (2011). *Tourism Satellite Account 2011*. Available at: http://www.ausstats.abs.gov.au/ausstats/subscriber.nsf/0/3E281CC5A71E3F91CA25796C00143363/$File/52490_2010-11.pdf [accessed: 13 April 2012].

BITRE (2006). *Aviation Statistics: Airport Traffic Data 1995–96 to 2005–06*. Canberra, Australia: Australian Government Bureau of Infrastructure, Transport and Regional Economics

BITRE (2008). *Aviation Statistics: Airport Traffic Data 1997–98 to 2007–08*. Canberra, Australia: Australian Government Bureau of Infrastructure,

Transport and Regional Economics. Available at: http://www.bitre.gov.au/publications/90/Files/AirportPublication2007-08.pdf [accessed: 6 February 2009].

BITRE (2011). *Statistical Report – Aviation*. Canberra, Australia: Australian Government Bureau of Infrastructure, Transport and Regional Economics.

Bjelicic, B. (2004). Osteuropa – Wachstumsmarkt für Low Cost Airlines, in *Internationales Verkehrswesen*, 56(7/8), 309–13.

Business Day (2009). Jetstar add extra planes to its Christchurch fleet. Available at: http://www.stuff.co.nz/4310735a13.html [accessed: 8 March 2009].

CAPA (2008). *Virgin Blue Launches 'Premium Economy' Product*. Sydney, Australia: Centre for Asia Pacific Aviation. Available at: http://peanuts.aero/low_cost_airline_news/airline/8126/59/Virgin+Blue+launches+%E2%80%98Premium+Economy%E2%80%99+product [accessed: 9 February 2009].

CAPA (2010). *Virgin Blue Rocks the Qantas Boat with Game Changing Moves: Etihad Deal, A330s in Domestic Market*. Sydney, Australia: Centre for Asia Pacific Aviation. Available at: http://www.centreforaviation.com/analysis/virgin-blue-rocks-the-qantas-boat-with-game-changing-moves-etihad-deal-a330s-in-domestic-market-33930 [accessed: 13 April 2012].

Doganis, R. (2001). *The Airline Business in the 21st Century*. London: Routledge.

Edwards, J. (2008). What is a low cost airline? Defining carrier business models, in *Proceedings of the Air Transport Research Society (ATRS)*, World Conference 6–10 July 2008, Athens, Greece.

Forsyth, P. (2003). Low-cost carriers in Australia: Experiences and impacts. *Journal of Air Transport Management*, 9, 277–84.

Forsyth, P. (2007). Predatory behaviour in Australian aviation markets, in Forsyth, P., Gillen, D.W., Mayer, O.G. and Niemeier, H.-M. (eds), *Competition versus Predation in Aviation Markets – A Survey of Experience in North America, Europe and Australia*. Aldershot, UK: Ashgate, 81–94.

Francis, G., Humphreys, I., Ison, S. and Aicken, M. (2006). Where next for low cost airlines? A spatial and temporal comparative study. *Journal of Transport Geography*, 14, 83–94.

Goettert, J.-M. and Schmidt, L. (2005). Irischer Geizhals. *FVW International*, 30, 66–9.

Gross, S. and Schröder, A. (2007). Basic business model of European low cost airlines – An analysis of typical characteristics, in Gross, S. and Schröder, A. (eds), *Handbook of Low Cost Airlines – Strategies, Business Processes and Market Environment*. Berlin: Erich Schmidt Verlag, 31–50.

Hazledine, T. (2008). Competition and competition policy in the trans-Tasman Air travel market. *The Australian Economic Review*, 41(4), 337–48.

InvestSMART (2009). *Virgin Blue Holdings Limited (VBA)*. Available at: https://www.investsmart.com.au/company_profile/summary/default.asp?SecurityID=VBA&ExchangeID=ASX [accessed: 11 February 2009].

New Zealand Herald (2010). Pacific Blue axing NZ domestic flights. Available at: http://www.nzherald.co.nz/travel/news/article.cfm?c_id=7&objectid=10666453 [accessed: 14 March 2012].

Pompl, W. (2007). *Luftverkehr – Eine ökonomische und politische Einführung*. Berlin: Springer-Verlag.

Qantas Airways (2007). Qantas announces profit result for year ended 30 June 2007 and an-market share buy-back, 16 August. Available at: http://www.jetstar.com/~/media/files/pdf/news/2007/aug/20070816.pdf [accessed: 4 February 2009].

Qantas Airways (2008). *Annual Report 2008*. Mascot, Australia: Qantas Airways. Available at: http://griffithaccountingandfinance.riverinainstitute.wikispaces.net/file/view/Qantas%2520Annual%2520Report%25202008.pdf [accessed: 25 March 2013].

Ramm, T. (2002). Germanisch fliegen. *Touristik Report*, 20, 10.

Statistics New Zealand (2009). *Civil Aviation*. Wellington, New Zealand: Statistics New Zealand Tatauranga Aotearoa. Available at: http://www2.stats.govt.nz/domino/external/web/nzstories.nsf/0/d94ee48e2b473672cc256b1f

Statistics New Zealand (2011). *Tourism Satellite Account: 2011*. Wellington, New Zealand: Statistics New Zealand Tatauranga Aotearoa. Available at: http://www.stats.govt.nz/~/media/Statistics/browse-categories/industry-sectors/tourism/tourism-satellite-account-2011/tsa-2011.pdf [accessed: 12 April 2012].

Statistics New Zealand (2012). *International Visitor Arrivals to New Zealand: January 2012*, Wellington, New Zealand: Statistics New Zealand Tatauranga Aotearoa.

Sterzenbach, R. and Conrady, R. (2003). *Luftverkehr – Betriebswirtschaftliches Lehr- und Handbuch*. Munich, Germany: Oldenbourg.

Tiger Airways (2009). *Company Overview*. Available at: http://www.tigerairways.com/au/en/about_us.php [accessed: 9 February 2009].

Tourism Australia (2008). *New Zealand Aviation Profile*. Available at: http://www.tourismaustralia.com/content/New%20Zealand/profiles_2008/Aviation%20profiles%202008-%20NZ%20[Compatibility%20Mode].pdf [accessed: 6 February 2009].

Tourism Australia (2010). *New Zealand Aviation Profile: Understanding how NZ Tourists Travel to Australia*. Available at: http://www.tourism.australia.com/en-au/documents/Corporate%20-%20Markets%20-%20New%20Zealand/Aviation_NZ_May10.pdf [accessed: 14 March 2012].

Tourism Australia (2011). *Quarterly Market Update – September 2011 Quarterly Results of the International Visitor Survey*. Sydney, Australia: Tourism Australia. Available at: http://www.tourism.australia.com [accessed: 23 March 2013].

Tourism New Zealand (2008). *Australia Market Overview*. Available at: http://www.tourismnewzealand.com/tourism_info/market-research/market-guides/australia/market-snapshot-australia.cfm [accessed: 6 February 2009].

Tourism New Zealand (2009). *New Zealand Airports*. Available at: http://www.newzealand.com/travel/getting-to-around-nz/getting-to-nz/airports/airports_home.cfm [accessed: 6 February 2009].

Virgin Blue (2006). *Hawaiian Airlines And Virgin Blue Make Hawaii Travel Easier From All of Australia*. Available at: http://www.virginblue.com.au/ AboutUs/Media/NewsandPressReleases/U_001186.htm [accessed: 9 February 2009].

PART V
Africa and the Middle East

Chapter 10
Low Cost Carriers in Africa

Uwe P. Hermann and Marius Potgieter

Introduction

Africa is the second largest continent in the world in terms of area – it extends from 37°21' N to 34°51'15" S over a distance of 8,000 kilometres – and of population. The continent consists of 54 sovereign states and includes various offshore islands. Africa is divided into four economic regions for the purpose of this chapter: Northern (countries north of the Sahara); Western (UEMOA); Eastern (COMESA); and Southern Africa (SADC). Aviation, especially low cost carriers (LCCs), will be dealt with in terms of these African regions. This chapter commences with the aviation profile of Africa and this is followed by a discussion of the African aviation industry and the LCC market. The chapter concludes with a presentation of two 'best practice' examples of LCC airlines, one in the north and one in the south.

Profile of Aviation in Africa

Africa is the world's least developed continent and this influences its aviation industry. The need for development and the funding of infrastructure still limits the progress of Africa's aviation industry The International Federation of Air Line Pilots' Associations (IFALPA) said in the late 1990s that at least 75 per cent of the air traffic infrastructure in Africa 'is unable to provide the services necessary for the safe and expeditious operations of flights' (Pawson 1997). This was followed by the statement that 'in 2004, a quarter of all accidents occurred in Africa – a region where air traffic is growing rapidly' (BBC News 2005) A US Government Accountability Office reported that 'only four countries in all of Africa have airport and airline safety standards meeting the requirements to receive the most favorable rating by the Federal Aviation Administration' (Hughes 2009). However, Logan and Demand Media (n.d.) cite Africa's accident rate as '9.2 per million planes taking off in 2005, 9.94 in 2009 and 7.41 in 2010', indicating an improvement in the air safety over Africa.

The enormity of the African continent necessitates a well-developed aviation industry. Commercial long haul flights to various African destinations were initiated during the colonial era (*Africa Travel Magazine* n.d.) to link European countries with their former African colonies. European carriers developed and maintained only the infrastructure they needed, leaving behind a deficiency of inter- and intra-regional and organizational cooperation between African countries and territories, as well as a lack of aviation industry expertise. This resulted in sporadic developments with hardly any integration between countries and regions after independence. Aviation and the facilities of many international airports in Africa are being expanded and upgraded to form a network of 'air bridges' (Africa Point and Travel Mall Africa n.d.).

Aviation in Africa

Aviation in Africa develops at an accelerating rate and there are currently 246 airlines holding an Air Operator Certificate. The focus of this discussion is on passenger services and does not include the impact of freight services. The ten largest passenger airlines in Africa for 2010 were: EgyptAir; South African Airways; Royal Air Maroc; Comair (including Kulula); Air Algérie; Ethiopian Airlines; Kenya Airways; Tunisair, Arik Air; and Air Mauritius (ATW 2011). There are about 4,000 recorded airports, airfields, and landing strips and less than 20 per cent have paved runways. The busiest airports on the continent, in terms of passenger traffic, are indicated in Table 10.1.

Africa's air transportation market grew by 2.1 per cent, and the 2010 forecasted Passenger Kilometres Performed (PKPs) for Africa is 7.6 per cent. However, 2008

Table 10.1 Busiest airports in Africa (2010)

No.	Country	City	Airport	Total passenger movements (millions)
1	South Africa	Johannesburg	OR Tambo International Airport	18.38m
2	Egypt	Cairo	Cairo International Airport	16.15m
3	Egypt	Sharm el-Sheikh	Sharm el-Sheikh International Airport	8.69m
4	South Africa	Cape Town	Cape Town International Airport	8.11m
5	Egypt	Hurghada	Hurghada International Airport	8.06m
6	Morocco	Casablanca	Mohammed V International Airport	7.25m
7	Nigeria	Lagos	Murtala Muhammed International Airport	6.30m
8	Kenya	Nairobi	Jomo Kenyatta International Airport	5.48m
9	South Africa	Durban	King Shaka International Airport	4.75m
10	Tunisia	Tunis	Carthage Airport	4.60m

Source: Compiled from various websites (2011).

was declared the year with the slowest rate of growth for the air transportation industry since 2002 by the International Civil Aviation Organization (ICAO 2009); at the same time, it also indicated that LCCs in Europe experienced double-digit growth rates. The forecast for Africa was adapted to only a 7 per cent PKP for 2010.

Air transportation in Northern Africa developed concurrently with aviation in Europe due to the close proximity of these African countries to Europe. The development of aviation in Northern Africa was a purposeful and integrated cohesion between regional governments. However, various LCCs from other continents still operate flights to mostly Northern Africa, such as Atlas Blue.com, Clickair, Helvetic, Transavia, Corendon, and Ryanair.

Cairo International Airport (Egypt) is the gateway to northern Africa and it is the second busiest airport in Africa, as indicated in Table 10.1 (Davitt 2011). It is the hub of EgyptAir, Africa's largest airline, which operates flights to 79 destinations. EgyptAir became a Star Alliance member in July 2008 and this increases Cairo International Airport's potential to become the largest hub in Africa (Africa Point and Travel Mall Africa n.d.). The Mohammed V International Airport in Casablanca (Morocco) is the sixth busiest airport in Africa and also the hub of Royal Air Maroc, which operates to 80 destinations, the highest for any of the airlines in Africa. Monastir Airport in Tunisia is the tenth largest airport, from which Tunisair (the eighth largest airline in Africa) operates to 33 destinations.

Ten of the countries in Western Africa signed the Treaty on Air Transport in Africa on 28 March 1961 and this led to the establishment of Air Afrique, which was to be the only official transnational carrier for West Africa (MyFundi.co.za n.d.). Air Afrique extended its operations and operated flights to various European destinations, but went bankrupt in 2001. The Murtala Mohammed International Airport in Lagos (Nigeria) is the busiest international airport in Western Africa and it is also the seventh busiest airport in Africa.

Florence Wilson, the pioneer of aviation in Eastern Africa, operated Wilson Airways, which was taken over by the Royal Air Force on the outbreak of World War II. The East African High Commission was established in 1948 and was replaced by the East African Common Services Organization in 1961. One of the departments of this organization was the East African Directorate of Civil Aviation, and the national airline of Kenya, Uganda and Tanganyika (Tanzania), the East African Airways Corporation (EAA), operated scheduled and charter flights (Flight International 1975 and MyFundi.co.za n.d.). A consequence of independence was the negotiations and renegotiation of bilateral air services agreements (ASAs), including traffic rights to and through East Africa. EAA was one of only a few airlines in the world that operated at a profit, but it was dissolved in 1977 when the former member countries established their own national carriers.

The Jomo Kenyatta International Airport in Nairobi (Kenya) provides easy access to Eastern and Central Africa and is the eighth busiest airport in Africa. This airport is the hub of Kenya Airways, the seventh largest airline in Africa, operating flights to 54 destinations. Ethiopian Airlines is the sixth largest airline in

Africa and operates one of the most extensive networks in Africa, which includes 76 destinations.

Aviation in Southern Africa emerged when Major Miller established Union Airways in 1929; this merged with South West African Airways to become Union Airways in 1932, the first commercial airline in this region (MyFundi.co.za n.d.). The assets of Union Airways were taken over by the South African government and renamed South African Airways (SAA). SAA is administered by TRANSNET, a government parastatal, making SAA one of only a few national carriers that is still 100 per cent owned by national government. SAA is today one of the oldest airlines in the world and also the first commercial airline in this region of Africa. SAA started flights into the rest of Africa and also international flights in 1945 to Bournemouth in the UK (a DC-4 Skymaster flight took 35 hours with a capacity of 44 passengers) after establishing a partnership with the British Overseas Airways Corporation (BOAC) (SAPFA n.d.). SAA was the first non-British airline in the world to fly a passenger jetliner (the De Havilland Comet) in 1953. SAA's first non-stop flight between Johannesburg and London occurred in 1983 with its acquisition of a Boeing 747-300 SUD. SAA is the second largest airline in Africa and operates flights to 35 destinations from Johannesburg. SAA was named the best airline in Africa at the 2011 World Airline Awards held in Paris (Skytrax n.d.) and SAA was also rated the safest airline in Africa in 2010 (McLachlan 2010).

Numerous private airlines attempted to operate alongside SAA but without success. Comair (Commercial Air Services) did take on the dominance of SAA. Comair is a private airline and was founded in Johannesburg in 1943. It was only five years later that they began scheduled air services between Johannesburg and Durban. Comair introduced Boeing 737-200 aircraft in the 1990s in direct competition to SAA. Comair entered into a franchise agreement with British Airways and rebranded the airline BA/Comair in 2000. Comair is the second largest airline in South Africa and provides full fare services throughout Southern Africa.

Another major airline in South Africa is Airlink, which provides a feeder service from/to smaller towns in South Africa. Airlink can be traced back to 1967, but officially started operations in 1995. Airlink has been in a strategic alliance with SAA and SA Express Airways since 1997. South African Express Airways (SA Express) started operations in April 1994 as a feeder airline for SAA. SA Express operates independently from SAA, although in a strategic alliance with SAA and Airlink.

Other Southern African countries also established their own airlines. For example; Air Rhodesia, a subsidiary of Central African Airways (which lasted only three years), started to operate commercial flights in 1964 and was then renamed Air Zimbabwe (MyFundi.co.za n.d.).

The regional office for Africa and the Indian Ocean of the International Air Transportation Association (IATA) is based in Johannesburg (South Africa) and their main activities in the African region are: safety; environment; operations and infrastructure; simplifying the business; financial services; and, lastly, industry charges, fuel and taxation (IATA 2011a).

Airports in South Africa

Southern Africa is easily accessible through South Africa. OR Tambo International Airport is the busiest hub in Africa and its duty free shopping complex is second in size only to that of Heathrow Airport in London (Africa Point and Travel Mall Africa n.d.). Cape Town International Airport is the second largest (fourth largest in Africa), and the King Shaka International Airport (Durban) is the third largest (ninth in Africa).

Nine of the major Airports in South Africa have been operated by the Airports Company South Africa (ACSA) since July 1993, a government parastatal. The major airports in South Africa are hubs for airlines, including LCCs. Smaller airlines are predominantly based at secondary and smaller airports and airfields. Very few of the airports in South Africa are operated as commercial entities or hold international status.

Unlike airports in Europe, airports in South Africa are not marketed as LCC hubs as these carriers predominantly operate on higher density routes between the major airports. A recent development is the utilization of secondary airports in order to ease congestion and costs associated with major airports.

Lanseria International Airport (generally known as 'Lanseria'), north-west of Johannesburg, was established in 1974 as a private airport and serves as a base for numerous charter airlines. Kulula Air, through its holding company Comair, collaboratively funded the expansion of Lanseria in 2006 to provide for scheduled air traffic and Lanseria obtained international status. This agreement entitled Kulula Air the exclusive right to operate scheduled LCC flights from Lanseria. The exclusive agreement of five years expired in March 2011 and 1time Airline indicated its intention to start operating from Lanseria and taking Kulula Air to the South African Competition Tribunal for unfair competitive practices. Mango started operating flights from Lanseria Airport on 1 June 2011. This airport is undoubtedly becoming a LCC hub in Johannesburg.

Wonderboom Airport is the only airport in South Africa owned by a municipality and it is situated in Pretoria, approximately 80 kilometres north of OR Tambo International Airport. Wonderboom Airport is in the process of upgrading its facilities and infrastructure and it is expected that scheduled services will operate from there in the near future. With increasing congestion at OR Tambo International Airport and LCC developments at Lanseria, it may be possible that Wonderboom Airport could in future be considered as a LCC hub for Pretoria.

The major airports in South Africa catering for LCC operations are shown in Table 10.2.

Table 10.2 South African airports from which low cost carriers operate

City	Airport	Departing passengers (2010)
Johannesburg	OR Tambo International Airport	9,772,000
Cape Town	Cape Town International Airport	4,236,000
Durban	King Shaka International Airport	2,411,000
Johannesburg	Lanseria International Airport	1,000,000+
Port Elizabeth	Port Elizabeth International Airport	760,000
East London	East London Airport	371,000
George	George Airport	336,000
Bloemfontein	Bloemfontein Airport	207,000
Nelspruit	Kruger-Mpumalanga International Airport (KMIA)	119,000

Sources: KMIA 2011, ACSA 2010, Venter 2011.

South African Travel and Tourism

South Africa has emerged as a major role-player in the domestic, regional, continental, and international tourism industry. The country has seen significant economic growth and international recognition since the advent of democracy in the 1990s and is still reaping the benefits of an escalating incoming international tourism industry, as indicated in Table 10.3.

The World Travel and Tourism Council (WTTC) predicts that tourism will contribute 9.6 per cent to the world's gross domestic product (GDP) and will employ over 120 million people by 2021 (WTTC 2011). It is further predicted that tourism will contribute 11.5 per cent of the GDP and employ over 1.7 million people by 2021 in South Africa (WTTC 2011).

International tourist arrivals to South Africa recorded an average growth rate of between 3 and 4 per cent per annum over the years 2010 to 2011 (SATourism, 2011). The domestic tourism market experienced a small but steady decline over the past five years. The South African Department of Tourism recently launched

Table 10.3 Tourism statistics for South Africa

Year	2005	2006	2007	2008	2009	2010	2011*	2012*
International arrivals** (in millions)	7.368	8.395	9.090	9.591	9.934	11.303	11.925	12.581
Domestic tourist trips	36.2	37.0	35.9	32.9	30.3	29.7	N/A	N/A

Notes: * Based on government predicted tourism growth of 4–7%, mean 5.5%; ** Figures represent total tourism arrivals (including same-day travellers and overnight visitors).

Source: South African Tourism (2010).

Table 10.4 Travel time (hours) and fares between Johannesburg and other major South African cities

Destination	Air	Rail	Coach/bus
Fare			
Johannesburg to Cape Town	Legacy = *R1,310.00 LCC = R669.99	R520.00	R575.00
Travel Time (hours)			
Cape Town	1.45	26.00	18.30
Durban	1.05	11.00	9.10
Port Elizabeth	1.40	20.00	16.15
East London	1.30	23.00	13.00

Note: *South African Rand (US$1 = ZAR 8).

Sources: SAA 2011, Greyhound 2011, Shosholoza 2011.

a campaign dubbed 'Sho't Left' (informal South African vernacular meaning 'a fast easy journey') in order to market the country among South Africans in an endeavor to increase domestic tourism. This campaign is a partnership between the public and the private sector and significant role-players in this campaign are airlines, especially LCCs.

Land transportation, such as private cars and long distance coaches, are the most popular modes of transport because of their relative low cost. However, travel time remains a major challenge because it takes about 14 to 16 hours to travel by road between Cape Town and Johannesburg. Flying between cities has for a long time been accessible only to business travellers or travellers who are not price-sensitive. Travel time between South Africa's three major cities from Johannesburg is shown in Table 10.4.

Air transportation is the most efficient form of transport in a country the size of South Africa and the emergence of LCCs in South Africa has made it possible for a greater number of people to travel within a shorter period of time at a reasonable cost. LCCs in South Africa experience significant competition from rail and road transport in terms of price; this is primarily as a result of high airport taxes charged at ACSA airports in South Africa. In some cases airport charges make up 50 per cent of the LCC airfare charged to passengers in South Africa (Billmore and Witepski 2011). ACSA airport taxes have increased by 70 per cent over the last two years, with an average annual increase of 5 per cent approved by government until 2014 (Brophy 2011). These high charges are hampering the development of LCCs in the country. LCCs are not the most cost-effective form of travel, but offer time saving as an added advantage, making them a popular choice.

The African Aviation Industry

Low cost carriers started to emerge in Africa after structural changes in Africa's aviation industry took place. African colonies were served by European airlines and a period of protectionism and a bilateralism approach followed independence, when countries prided themselves on their national flag carriers while protecting their markets through restrictions and regulation. National airlines were government owned and managed and many of these airlines did not survive. The United Nations Economic Commission for Africa (UNECA n.d.) describes the characteristics of Africa's aviation industry as:

- interference of policy makers and governments; more or less 53 non-physical barriers to access air transport markets in Africa;
- the lack of cooperation among airlines and air space regulatory authorities; this includes limited collaboration amongst African carriers relating to travel through the continent, such as when travelling between two capital cities; and,
- inadequate air transport policy (there were 11 regulation policies in 1990).

African countries were resistant to the introduction of change while the rest of the world introduced liberalization, deregulation, alliances and multilateral regulations. Africa's aviation industry cannot develop in isolation because globalization calls for appropriate reforms. The lack of acceptable policies and regulations remains one of the major barriers to the expansion of Africa's aviation industry.

Regional initiatives were developed and adopted in Africa, for example the Yamoussoukro Declaration of 1988 and the Yamoussoukro Decision of 1999. Various other sub-regional initiatives were initiated by inter alia COMESA (Common Market for Eastern and Southern Africa), ECOWAS (Economic Community Of West African States), and SATCC (Southern African Transport and Communications Commission).

Yamoussoukro Declaration

This declaration is based on a new African air transport policy (integrating airlines and the establishment of regional controlling bodies) and was signed at the Conference of African Ministers Responsible for Civil Aviation in November 1988 in Yamoussoukro (Côte d'Ivoire) (UNECA n.d. and MyFundi.co.za n.d.). Signatories did not consider the privatization of their national flag carriers at this stage. This was followed by a meeting of the African Ministers of Transport and Communications who met in Mauritius on 9 September 1994 (UNECA 2009a) and some of the main resolutions were:

- To incorporate the Declaration within one year;

- To implement flexibility in exchange and liberalize traffic rights;
- The establishment of multinational airlines; and
- Cooperation and integration programmes.

A reluctance to implement this policy prompted the Banjul Accord, signed in 1996. The Yamoussoukro Declaration was reviewed and it was discovered that the main reason why the implementation date of 1996 was not realized was the issue of traffic rights.

The economic environment of Africa underwent major changes and African countries started to embrace private sector participation, such as LCCs, and this called for a review of the Yamoussoukro Declaration.

Yamoussoukro Decision

The Yamoussoukro Decision (YD), intended to be Africa's 'Open Skies Policy', was signed in 1999, by all African countries except for the Arab Maghreb Union (AMU) countries (Algeria, Morocco, Tunisia, Libya and Mauritania) (UNECA 2010). A consequence of this was the establishment of the African Economic Community and the gradual liberalization of intra-African air transportation (particularly the fifth freedom traffic rights) (UNECA n.d. and MyFundi.co.za n.d.). A summit was held in July 2000 in Lomé (Togo) where the heads of state and government of the African Union reviewed the progress in the implementation of the Yamoussoukro Declaration and reaffirmed the commitment of African countries to gradually eliminate non-physical barriers to intra-Africa air transport (UNECA 2009a).

The next step was to regulate aviation between Africa and the rest of the world and a direct result of the Decision was that ten African countries would sign open skies agreements with the USA by the end of 2001 (MyFundi.co.za n.d.). African Ministers responsible for Air Transport have since held three meetings (2005 at Sun City, South Africa; 2006 in Libreville, Gabon; and 2007 in Addis Ababa, Ethiopia) and decided to create the Executing Agency of the Yamoussoukro Decision and to entrust it to the African Civil Aviation Commission (AFCAC). Escalating freedom of the African skies brings about new opportunities and Abdoulie Janneh, the UN Under-Secretary-General and Executive Secretary of the Economic Commission for Africa, said in the opening speech at the High Level Meeting of African Airline Companies held in May 2006 in Tunis (Tunisia) that: 'Over the past fifty years, despite the several constraints, air transport has steadily brought African countries closer together, linking most African capital cities to the rest of the continent by air. It has also contributed to the expansion and deepening of intra-African commerce and trade' (Janneh 2006). The ministers responsible for civil aviation in Central and Western African countries held a meeting on 7 November 2008 in Accra (Ghana) and agreed to accelerate the liberalization of the markets in their sub-regions (UNECA 2009b).

The New Partnership for Africa's Development (NEPAD) (UNECA 2010) recently surveyed the implementation of the Yamoussoukro Decision at a regional level and concluded the following:

- *Northern Africa*: A draft convention to liberalize air transport is being prepared, but neither a liberalization agreement nor any agreement in conformity has been concluded.
- *Western Africa*: All eight West African Economic and Monetary Union (UEMOA) member states and members of the Banjul Accord Group (BAG) have fully complied with the decision.
- *Central Africa*: Only the member states of CEMAC (Economic Community of Central African States) and BAG are in conformity with the decision.
- *Eastern and Southern Africa*: COMESA, the East African Community (EAC) and the Southern African Development Community (SADC) jointly adopted the COMESA–EAC–SADC Competitive Regulations on Air Transport Services liberalization, Provisions and Procedures for the implementation of common regulation in 2008. A joint competitive authority (JCA) was established in 2008 to oversee the air transportation liberalization process.

African governments are urged to effectively implement the African policy on liberalization in order to enhance the intra-African aviation trade and tourism through regional cooperation. South Africa has embarked on the review of bilateral air services arrangements in line with the YD principles' and calls on all other African states to liberalize air transport almost immediately (ICAO 2010).

Changes in the Consumer Market

Low cost carriers started a new trend and currently hold a 24 per cent market share worldwide (IATA 2010). Air travel markets have declined since February 2011 and this is ascribed to the earthquake and tsunami in Japan, as well as the political unrest in northern Africa which seriously disrupted air traffic in Egypt and Tunisia. Added to this is the increase in oil prices to US$120 a barrel, which results in higher fares, and this has an adverse effect on LCCs. It is concerning that the Passenger Load Factor (PLF) for Africa declined from 68.2 for March 2010 to 62.7 for March 2011. A year-on-year comparison for passengers and freight is shown in Table 10.5.

Demand for air transport, measured by revenue passenger kilometres (rpk), are now 6 per cent larger than the pre-recession peak and the world GDP is expected to increase by 3.1 per cent in 2011 and passenger markets by 5.6 per cent, according to IATA (2011c), which also indicates that:

> African airlines should continue to benefit from the strong economic growth in these regions, albeit slower than in 2010 as central banks attempt to suppress an emerging inflation.

Table 10.5 Key passenger and freight figures (March 2011 vs March 2010)

Region	rpk	ASK	PLF	FTK	AFTK	FLF
Africa	-7.0%	1.0%	62.7	-2.8%	-1,8%	32.2

Note: rpk: revenue passenger kilometres; ASK: Available Seat Kilometres; PLF: Passenger Load Factor; FTK: Freight Tonne Kilometres; AFTK: Available Freight Tonne Kilometres; FLF: Freight Load Factor
Source: IATA 2011b.

Passengers are generally price-sensitive and this causes inter-model substitution, where passengers will rather switch to rail and coach transport if they feel that the price of a flight ticket is too expensive. This is a major threat to LCCs, which operate short haul flights. Policy makers and airlines, especially LCCs, need to find a balance between their fares (including all forms of taxes) and their income by considering the elasticity of demand when analysing policy proposals, such as liberalization, airport charges, and taxation or emission schemes.

The Low Cost Carrier Market

Low cost carriers entered the aviation market with the foundation of Southwest Airlines in 1971, the 'big brother' of LCCs. LCCs have held their position for a

Figure 10.1 South African Airways created Mango to defend their markets against the encroachment of low cost rivals
Source: Michael Lück.

number of years in North America and Europe. Legacy carriers tried to fight LCC advances and were even beating them in the early 1980s (Graham and Vowles 2006) only to experience resurgence. Some legacy carriers followed the concept of developing carriers within carriers (CWC) as described in Chapter 1. This concept involves the creation of secondary LCC operations within the legacy carriers' operations. The main reasons for creating CWCs are firstly, to gain advantage from cost optimization; and secondly, to defend traditional markets against the encroachment of low cost rivals (Graham and Vowles 2006). SAA and Comair followed such a management strategy with the creation of Mango and Kulula Air (see Figure 10.1). Both these LCCs operate with the backing of their parent company in terms of infrastructure, aircraft and training.

Some LCCs venture further than their domestic services and also enter the long haul market through airline partnerships. The challenge for these airlines is to continuously search for innovative ideas that will stimulate demand, adaptation, and flexibility. LCCs contribute towards accelerated traffic growth although they are confined to regions, for example northern and Southern Africa.

Low Cost Carriers in Africa

The deregulation and liberalization of the African skies brought about new opportunities. LCC operations in Africa are predominantly located in Northern Africa (Egypt, Tunisia and Morocco) and Southern Africa (South Africa) and following is a synopsis of these LCCs as indicated in Table 10.6.

LCCs in northern Africa, serve domestic routes, as shown in Table 10.6, but most of their flights are intercontinental. Europe is viewed as their most important market due to their proximity to all the airports of this continent.

Aero Contractors is an airline based in Nigeria, in Western Africa. This airline does not promote itself as a LCC but does exhibit some of the characteristics of a LCC operator and is included as an African LCC. Aero Contractors' flights can be booked on the Internet, fares include hidden fees, in-flight meals can be purchased, its staff is minimally uniformed, the company shares office space with other airlines, tickets are non-refundable and non-transferable, the airline does not have a frequent flyer programme, and fares are much lower than those of its competitors (Ajou 2010).

Eastern Africa has only one LCC, Fly540, in Kenya. Fly540 which began flying in 2006, operates from Nairobi (Kenya) with a fleet of ten aircraft. Fly540 and easyJet recently formed a new partnership and the intention is to launch FastJet, a new LCC which will eventually replace the Fly540 brand. FastJet is likely to operate with Airbus 319s or Embraer 190s (Rivers 2012) and is currently seeking growth throughout the East African region as well as southern Africa (South Africa).

Table 10.6 Low cost airlines in Africa (1991–2011)

Name	Country	Year of foundation	Start of LCC operations	Withdrawal from market	Parent/holding company
Northern Africa					
AtlasBlue	Morocco	2004	2004	2010 (taken over by Royal Air Maroc)	
Air Arabia Egypt	Egypt	2009	2010	—	Air Arabia (40%) Travco Group (50%)
Air Arabia Maroc	Morocco	2009	2009	—	Air Arabia
Jet4You	Morocco	2006	2006	—	TUI Travel PLC
Karthago Airlines	Tunisia	2001	2002	—	Karthago Group
Western and Central Africa					
Aero Contractors	Nigeria	1959	1960	—	
Elysian Airlines	Cameroon	2006	2006	2010	
Eastern Africa					
Fly540	Kenya	2006	2006	—	Lonrho PLC (UK)
Southern Africa					
Air Kumba	Zimbabwe	2009	2009	2010	Air Zimbabwe
Fly540 Angola	Angola	2010	2011	—	Lonrho PLC (UK)
Intensive Air	South Africa	1993	2000	2002	
Interlink Airlines	South Africa	1998	2010	2010	
Kulula Air	South Africa	2001	2001	—	Comair Limited
Mango	South Africa	2006	2006	—	SAA
Nationwide Airlines	South Africa	1991	2006	2008	
SANTACO Airlines	South Africa	2011	2011	—	SANTACO
Velvet Sky	South Africa	2009	2011	2012	Macdonald Holdings
1Time	South Africa	2003	2004	—	1Time Holdings

Sources: Compiled from various websites (May–October 2011).

LCCs are well represented in Southern Africa, specifically South Africa. However, the protection of SAA hampers the rapid expansion of private airlines and it took almost a decade for a LCC to take to the South African skies after aviation deregulation.

SAA was the only South African international airline until the founding of Trek Airways in 1953. Trek Airways operated low cost non-scheduled flights to European destinations and Perth (Australia). The airline became very popular, branding itself a low fare airline, although not a LCC according to the criteria indicated by Harbison and McDermott (2009) and it is therefore not included in Table 10.6, however, it requires noting. Guttery (1998) indicates that Trek Airways charged very low fares compared to their competitors. South Africa slipped into international isolation in the 1960s during the period of segregation and SAA, the national carrier of South Africa, was prohibited from flying over Africa. This seriously hampered its operations. Trek Airways, with its close connection to Luxair, re-registered its aircraft in Luxembourg and this exempted it from African overflight restrictions (aircraft re-registered in Luxembourg flew under the name of Luxavia). With the lifting of international sanctions and the deregulation of South Africa's aviation in the 1990s, Trek Airways adapted its strategy to remain competitive with SAA and started a subsidiary named Flightstar, which operated the first fleet of A320 aircraft in Africa. Flightstar utilized vast start-up funds from its parent company, which resulted in the company ceasing all operations in 1994 and this brought about the end of Trek Airways.

Contemporary LCCs emerged over time in South Africa when deregulation made it possible to establish new airlines and to expand existing ones. The development of LCCs in South Africa began with Intensive Air. This airline diversified its services and started low cost flights between Johannesburg and Cape Town in 2000 and terminated its services in 2002. Comair, in 2001, launched South Africa's first still-existing LCC, namely Kulula Air. Soon thereafter other LCCs followed: 1time Airline and Mango.

A latest addition to South Africa's portfolio of LCCs is SANTACO Airlines. This LCC came into existence in mid-2011 and is owned by SANTACO (South African National Taxi Council) and its flights are operated by AirQuarius Aviation. SANTACO represents taxi operators throughout South Africa and its ambition is to make air travel affordable to the general South African public (SANTACO 2011). SANTACO Airlines operates flights from Lanseria to Cape Town and Bhisho.

In other cases in South Africa some legacy carriers such as Nationwide Airlines and Interlink Airlines later developed LCC characteristics with their low fares, no frills and high-density layouts. These LCCs offered services in economy class while providing business class service on the same flights. This strategy was adopted in order not to compete directly with SAA but rather to compete within the growing LCC market. Franke (2004) states that this business strategy is the most viable strategy for network carriers as it creates separate business systems for distinct customer segments. In the case of Nationwide and Interlink, these

Table 10.7 Fleet structure of African low cost carriers

Name	Number of aircraft	Number and type of aircraft	Different types	Orders
Northern Africa				
Air Arabia Egypt	3	3 Airbus A320	1 type	1 Airbus A320
Air Arabia Maroc	5	5 Airbus A320	1 type	5 Airbus A320
Jet4You	9	6 Boeing B737-400 3 Boeing B737-800	1 type with 2 subtypes	
Karthago Airlines	1	1 Boeing B737-300	1 type	
Western and Central Africa				
Aero Contractors	10	2 Boeing B737-400 5 Boeing B737-500 3 Bombardier Dash 8-300	2 types with 2 subtypes	2 Boeing 737-500
Eastern Africa				
Fly540 Kenya	10	3 Beechcraft 1900 1 Bombardier CRJ-100 3 De Havilland Canada DHC Dash 8 1 Fokker 28 2 Cessna 208	5 types	9 ATR72-500
Southern Africa				
Fly540 Angola	2	1 ATR 42-300 1 ATR 72-500	1 type with 2 subtypes	
Kulula Air	10	1 Boeing B737-300 5 Boeing B737-400 4 Boeing B737-800	1 type with 3 subtypes	8 Boeing 737-800
Mango	4	4 Boeing 737-800	1 type	Undisclosed number of A320 family
SANTACO Airlines	1	Boeing B737-200	1 type	
Velvet Sky	3	3 Boeing B737-300	1 type	
1time Airline	11	1 MD-81 2 MD-82 5 MD-83 3 MD-87	1 type with 4 subtypes	

Sources: Compiled from various websites (May 2011).

airlines offered one business stream catering for two distinct customer segments. This management model has not proven feasible because both these carriers failed to remain competitive and both terminated their operations soon after offering LCC services.

Low Cost Carrier Aircraft Fleet

LCCs in South Africa lease their aircraft in order to curb costs. The initial equipment of all LCC operators in South Africa consisted of used aircraft and the operators' ability to offer low fares can be attributed to the aircraft longevity (an average of 20 years). These aircraft have higher operating costs, but they are relatively affordable to source. Mango and Kulula Air started their operations with equipment obtained from their parent airlines (SAA and Comair). The ability for these parent carriers to place large orders with manufacturers allows cost advantages to flow to their associated LCCs. Mango is due to replace its fleet of Boeing 737-800s with the Airbus A320 family and this is in line with the parent company's process of achieving fleet standardization. Comair has ordered a new fleet of Boeing 737-800s for Kulula Air. Although 1time Airline possesses the largest fleet of all the LCCs in South Africa, Kulula Air is considered the largest LCC, based on capacity.

Apart from Kenya's Fly540, LCCs in Africa generally have fleet uniformity. Fly540 is in the process of fleet renewal with a recent order for ATR72s and this should standardize their fleet. Fly540 is currently establishing subsidiaries in Tanzania and Ghana under the Fly540 brand, similar to that in Angola, and will have a fleet similar to Kenya's fleet. Other LCCs, such as Air Arabia (in Morocco and Egypt) placed Airbus orders. Table 10.7, on the previous page, shows the fleet structure of LCCs in Africa.

Best Practices

Two outstanding achievers in the LCC industry in Africa are firstly, Air Arabia in northern Africa; and, secondly, Kululu Air in Southern Africa. These two LCCs are opposites, one in the north versus one in the south. Air Arabia in the north enjoys easy access to the vast European and Middle Eastern markets and hubs, while Kulula Air on the southern tip of the African continent has to cope with the local and regional markets due to the long haul distances to other continents. Kulula Air dates back to 2001 and is currently the oldest operating LCC in Southern Africa, while Air Arabia dates back to 2003 and is one of the youngest LCCs in the northern part of the African continent (it commenced operations in Morocco in 2009 and in Egypt in 2010). Air Arabia further enjoys the backing of a major holding company, while Kulula Air does not have similar support.

Here follows a synopsis of these two African LCCs.

Air Arabia

Air Arabia has its head office in Sharjah (United Arab Emirates). The first and largest low cost carrier in the Middle East, it commenced services on 28 October 2003. The value proposition of Air Arabia is: 'Pay Less. Fly More'. Air Arabia offers connections with other flights – its own and those of other airlines – at its base airport; it operates services to 46 destinations. Air Arabia expanded its operations and established three subsidiaries, of which two are situated in northern Africa: Air Arabia Egypt and Air Arabia Maroc. The third subsidiary is Air Arabia Jordan in the Middle East. Air Arabia Maroc started operations on 6 May 2009 and operates LCC flights from Casablanca to 10 countries; and Air Arabia Egypt started operations on 1 June 2010 and operates LCC flights to five countries from Alexandria. A LCC agreement was also signed with Yeti Airlines in Nepal and flights were operated from this base until operations were suspended due to the uncertain political and economic situation in early 2008. Some of the awards Air Arabia has received over the years are: the Middle East and North Africa (MENA) travel award for best airline in 2005, 2006, 2007 and 2008; the Centre for Asia Pacific Aviation award for best low fare airline for 2006; the World Airline Award for best Low-Cost Airline in MENA by Skytrax in 2007, 2008 and 2009; the LCC of the year at the Aviation Business Awards in 2007, 2008 and 2009; and the World's best LCC by *Aviation Week* (Air Arabia 2011). Indeed an example of a very successful LCC.

Kulula Air

Kulula Air, also known as kulula.com or just Kulula, was founded in 2001 and has operated LCC flights since August 2001 from its Johannesburg base to nine destinations, including international flights to Namibia, Mauritius, Zambia and Zimbabwe. *Kulula* is a Zulu word meaning 'easy', depicting the simple nature

Figure 10.2 One of Kulula's aircraft is covered with a legend describing the different parts of the airplane in a humorous way – it is known as 'Flying 101'

Source: Morné Booij-Liewes.

of the airline's operations compared to traditional carriers. The Kulula website provides facilities for booking flights not only with Kulula Air but also on other airlines to selected destinations, as well as a range of travel-related services such as car hire and accommodation. Kulula is well known for its distinctive, brightly coloured and often humorous liveries and promotional material. One of its aircraft is covered with a legend describing the different parts of the aeroplane in a humorous way and it is known as 'Flying 101' (see Figure 10.2 on the previous page). It is interesting to note that no Kulula aircraft or any employees are ever depicted in Kulula advertisements. Kulula have a frequent flyer programme called 'Jetsetters'. The Kulula promise to its customers is:

> We promise to send you all the good stuff about our secret sales and juicy discounts. We won't bug you with any of the boring stuff, only great deals and super news.

The Kulula website also has humorous video clips, called 'Episodes', depicting the hilarious and brilliant years of the airline (see https://www.kulula.com/info/marketingmagic.aspx). Kulula's description of itself reads:

> We are more than just an airline … we're an entire experience.

Summary

Africa is a very large continent and the current state of its aviation industry is a confluence of various events and occurrences. The implementation of deregulation and liberalization brings new opportunities, albeit at a slow pace, for intra-regional and inter-continental aviation, as well as for LCCs. Added to this is facility and infrastructure development. The establishment of bilateral agreements and freedom of the African skies could lead to linking Africa's aviation from north to south and east to west, which will enable existing LCCs and other possible new entrants to provide needed, faster, and low cost transportation in, to, and from Africa. Conventional airlines and LCCs are investing in updating and standardizing their fleets and this will hopefully have a positive impact on the economic and natural environments in Africa. This industry provides ideal ground for research and closer cooperation between regions that could lead to the establishment of a knowledge base to the benefit of all interested in the aviation industry of Africa.

References

ACSA (2010). *Annual Airport Statistics*. Johannesberg, South Africa: Airports Company South Africa. Available at: http://www.airports.co.za/Tools/Documents/DocumentDownload.asp [accessed: 30 November 2011].

Africa Point and Travel Mall Africa (n.d.). *1. A Look at the Leading African Airline Hubs and Airports*. Available at: http://www.africapoint.com/newsletters/africa-airports.htm [accessed: 30 November 2011].

Africa Travel Magazine (n.d.). Transportation a key part of ATA Congress 2010 Agenda. Available at: http://www.africa-ata.org/page1.htm [accessed: 30 November 2011].

Air Arabia (2011). Get to know who we are, and how we are painting the world map red! Available at: http://www.airarabia.com/crp_1/awards&stitle=awards&pid=125? [accessed: 30 November 2011].

Ajou, O.D. (2010). *How Aero Contractors Offer Very Low Airfares: The Catches*. Available at: http://www.davidajao.com/blog/2010/03/08/how-aero-contractors-offers-very-low-airfares-the-catches/ [accessed: 30 November 2011].

ATW (2011). World Airline Traffic Results 2010, in *World Airline Report, Air Transport World*, 43–50. Available at: http://atwonline.com/sites/atwonline.com/files/misc/ATW-World-Airline-Report-2011.pdf [accessed: 11 July 2012].

BBC News (2005). How safe is it to fly in Africa? Available at: http://news.bbc.co.uk/2/hi/Africa/4521970.stm [accessed: 20 June 2012].

Billmore, S. and Witepski, L. (2011). Inbound industry 'desperate' says SATS as new air travel charges set to kick-in. *Travel Industry Review*, 197(8), 1–2.

Brophy, S. (2011). The great ACSA tax hike?, in GoTravel24 [gotravel.com, Online] 12 September. Available at: http://www.gotravel24.com/theme/feature-focus/great-acsa-tax-hike [accessed: 23 September 2011].

Davitt, D. (2011). *The Moodie Report: Cairo International Posts New Annual Record for Passenger Traffic*, 19 January 2011. Available at: http://www.moodiereport.com/document.php?c_id=6&doc_id=26275|The [accessed: 11 July 2012].

Harbison, P. and McDermott, P. (eds) (2009). *Global LCC Outlook Report*. Sydney, Australia: CAPA Centre for Asia Pacific Aviation, 266.

Flight International (1975). World airline directory, 20 March 1975. Sutton, UK: FlightGlobal, 484.

Franke, M. (2004). Competition between network carriers and low cost carriers – retreat, battle or breakthrough to a new level of efficiency? *Journal of Air Transport Management*, 10, 15–21.

Graham, B. and Vowles, T.M. (2006). Carriers within carriers: A strategic response to low-cost airline competition. *Transport Reviews*, 26, 105–26.

Greyhound (2011). Booking portal. Available at: www.greyhound.co.za [accessed: 14 September 2011].

Guttery, B.R. (1998). *Encyclopedia of African Airlines*. Jefferson, NC: McFarland and Co.

Hughes, D. (2009). Flying across Africa's skies. *ABC NEWS*, 6 October 2011. Available at: http://abcnews.go.com/Travel/flying-africas-skies-risky/story?id=8755299#.UAgt4PWKElo [accessed: 15 November 2011].

IATA (2010). *Low-cost Carriers*. Geneva: International Air Transport Association. Available at: http://www.iata.org/pressroom/airlines-internatinal/june-2010/Pages/ 10g.aspx [accessed: 14 September 2011].

IATA (2011a). *Africa*. Geneva: International Air Transport Association. Available at: http://www.iata.org/worldwide/africa/Pages/index.aspx [accessed: 4 November 2011].

IATA (2011b). *Air Transportation Market Analysis: March 2011*. Geneva: International Air Transport Association. Available at: http://www.iata.org/economics [accessed: 4 November 2011].

IATA (2011c). *Industry Financial Forecast: High Fuel Prices Squeeze Airline Profits*. Geneva: International Air Transport Association. Available at: http://www.iata.org/economics [accessed: 4 November 2011].

ICAO (2009). *Marginal Traffic Growth and Fuel Hedging Losses Take Toll on Airline Industry in 2008*. Montreal, Quebec: International Civil Aviation Organization, 5 June 2009.

ICAO (2010). *Liberalization of Air Services*. ICAO Assembly – 37th Session, Agenda Item 49: Liberalization of international air transport services, A37-WP/211, EC/13, 22 September 2010. Montreal, Quebec: International Civil Aviation Organization. Available at: http://www.sec. icao.int [accessed: 4 November 2011].

Janneh, A. (2006). *Strategies for the Development of the Airline Transport Industry in Africa*. Opening statement to High Level Meeting of African Airlines Companies, 29–30 May 2006. Available at: http://uneca.org/eca_resources/Speeches/Janneh/2006/290506 [accessed: 4 November 2011].

KMIA (2011). *Analysis – Passenger Statistics (TOTAL)*. South Africa: Kruger Mpumalanga International Airport. Available at: http://www.mceglobal.net/Portals/0/pax_apr2011.pdf [accessed: 4 November 2011].

Logan, G. and Demand Media (n.d.). *Air Travel Safety in Africa*. Available at: http://traveltips.usatoday.com/air-travel-safety-africa-35417.html [accessed: 20 June 2012].

McLachlan, S. (2010). *The Ten Safest African Airlines*, 4 May 2010. Available at: http://www.galding.com/2010/05/04/the-ten-safest-african-airlines [accessed: 11 July 2012].

MyFundi.co.za (n.d.). *Commercial Aviation in Africa I: The History of*. Available at: http://myfundi.co.za/e/The_History_of_Commercial_Aviation_in,Africa_I [accessed: 11 July 2012].

Pawson, L. (1997). *Africa's Deadly Skies (Dangers of Increased Air Traffic)*. Available at: http://www.highbeam.com/doc/1G1-19250861.html? [accessed: 20 June 2012].

Rivers, M. (2012). FastJet eyes late summer launch alongside Fly540 brand. *FlightGlobal News*. Available at: http://www.flightglobal.com/news/articles/fastjet-eyes-late-summer-launch-alongside-fly540-brand-373018/ [accessed: 15 June 2012].

SAA (2011). Booking portal. South African Airways website. Available at: http://www.flysaa.com [accessed: 14 September 2011].

SANTACO (2011). *Taxi commuters can now fly*. South African National Taxi Council. Available at: http://www.santaco.co.za/index.php/notice-board/item/72-taxi-commuters-can-now-fly [accessed: 3 December 2012].

SAPFA (n.d.). *History of Aviation in SA*. South African Power Flying Association. Available at: http://www.sapfa.org/history/history-aviation-sa [accessed: 11 July 2012].

Shosholoza (2011). *Shosholoza Meyl Fares and Schedules*. Available at: http://shosholoza-meyl.co.za/tourist_class_timetable.html [accessed: 14 September 2011].

SATourism. 2011. *Annual Report 2011*. Pretoria: South African Tourism.

Skytrax (n.d.). *South African Airways is Named the Best Airline in Africa for 2011*. Available at: http://www.worldairlineawards.com/Awards_2011/africa.htm [accessed: 11 July 2012].

UNECA (n.d.). *Civil Aviation Reform in Africa*. Addis Ababa: United Nations Economic Commission for Africa. Available at: http://www.uneca.org/itca/yamoussoukro/civil_aviation_africa.pdf [accessed: 11 July 2012].

UNECA (2009a). *The Africa Regional Review on Transport 2009*. Addis Ababa: United Nations Economic Commission for Africa

UNECA (2009b). *The Transport Situation in Africa*. Sixth session of the Committee on Trade, Regional Cooperation and Integration. Addis Ababa: United Nations Economic Commission for Africa.

UNECA (2010). *Assessing Regional Integration in Africa IV: Enhancing Intra-African Trade*. Addis Ababa: United Nations Economic Commission for Africa.

Venter, I. (2011). Lanseria weighs expansion options as passenger numbers swell. *Engineering News*, 4 February 2011. Available at: http://www.engineeringnews.co.za/article/airports-2011-02-04 [accessed: 4 November 2011].

WTTC (2011). *Economic Impact Data and Forecasts*. London: World Travel & Tourism Council. Available at: http://www.wttc.org/eng/Tourism_Research/Economic_Research [accessed: 11 July 2012].

Chapter 11
Low Cost Carriers in the Middle East

Andreas Wald

Introduction

One of the classic writings in strategic management distinguishes two fundamental ways to compete: companies should either differentiate their products and/or services by offering innovative products with superior quality and a high degree of customization, or they should aim at being a cost leader in the market. In the latter case customers may profit from low prices (Porter 1980).

Due to regulatory constraints, the airline industry was rather late in implementing a cost leadership strategy. The related business model has several names, for example, 'low cost carrier' or 'no frills airline', and there are many variations (Francis et al. 2006, Mason and Morrison 2008, Wensveen and Leick 2009, Koch 2010). The most important characteristic of the low cost business model is competition on price rather than on differentiated service. This competitive strategy requires a very strict cost optimization which goes far beyond the usual efforts of network carriers. A successful low cost carrier can realize up to 40–50 per cent lower per unit costs than an established network carrier (Franke 2004, Belobaba 2009, Koch 2010).

The low cost business model originated in the US, and in a second wave it became an established model in Western Europe. In a third wave, the low cost carrier (LCC) covered selected parts of East Asia and Southeast Asia (Francis et al. 2006 and Pitfield 2008, see also Chapter 1 of this volume). More recently, the low cost business model has also been adopted by airlines in the Middle East. The countries of this region differ considerably in terms of their size, population, structure of the economy, and development. As a consequence, the demand for air transport and the structure of the airline industry vary accordingly. The aim of this chapter, therefore, is to provide an overview of the market environment for LCCs in the Middle East. In contrast to other chapters of this book, which describe the status quo of the LCC in Western economies, special attention will be paid to the political, social and economic environment for air transport in Middle Eastern countries. The political instability in many parts of this region affects most kinds of businesses and will also have a significant impact on the future development of the aviation industry.

In the next part of this chapter, the main political, economic and socio-demographic factors of the different Middle Eastern countries will be described. There follows an assessment of the aviation industry in this region. The fourth part of the chapter portrays the status quo of the low cost airline industry. The focus of this chapter will be exclusively on LCCs registered in one of the countries of this region. A carrier registered in other parts of the world but serving the Middle East will not be considered.

The Middle East: Political, Social and Economic Factors

There is no clear cut definition of the geographical region which constitutes the Middle East. The 'Greater Middle East' encompasses North Africa, including countries like Sudan, Morocco and Somalia; and Western Asia, including countries like Kazakhstan, Pakistan, and Turkey (Goldschmidt and Davidson 2005). This chapter adopts a more narrow definition and considers only 15 countries stretching from the east of the Mediterranean to the Caspian Sea in the north, including the Arabian Peninsula in the south (AEA 2006) (see Table 11.1). The area covers a surface of 6.2 million square kilometres and hosts a population of approximately 290 million.

The similarities of Middle Eastern countries lie in Islamic culture and in the geographic conditions. Large parts of the territory are deserts or semi deserts. In addition to this, mountains make most parts of the area uninhabitable. Settlement is concentrated on the coasts and in fertile river valleys like the Nile, the Euphrates, the Tigris, and the River Jordan. The countries vary considerably in size and population. They include small states like Bahrain, as well as large and sparsely populated countries like Saudi Arabia. Table 11.1 provides fundamental economic, geographic, social and political data of the 15 Middle Eastern countries.

The Economy

The diversity of the Middle East is also reflected in the uneven distribution of wealth in and across the countries of this region. Whereas oil rich countries like Kuwait, Qatar and the United Arab Emirates are among the richest countries in the world, the Yemen, Egypt, the Palestinian territories and Syria are closer to the lower end of the scale. But even within the rich countries wealth is unevenly distributed. Factors of production, assets and political power are almost exclusively controlled by a small ruling class, leaving large parts of the population in poverty.

The industrial basis in Middle Eastern countries is broadly underdeveloped. The main sources of revenue are oil, gas and oil-related products for the countries of the Arabian Peninsula, Iraq and Iran. Oil importing countries like Jordan and the Lebanon have a more diversified economy. More recently, several oil producing countries started investing heavily in commercial, industrial and touristic infrastructure to reduce their dependence on a single commodity. Nonetheless, all industrial and infrastructure projects are dependent on cross-subsidization from the oil industry (Richards and Waterbury 2007). Besides the United Arab Emirates, tourism also plays an important role in Egypt, the most populated Middle Eastern

Table 11.1 Economic, geographic, social and political factors of the Middle East

Name	Population*	Area (km^2)*	GDP (US$)*	GNI per capita (US$)*	Human Development Index**	Democracy Index***
Bahrain	791,473	665	$20,594,899,946	$25,420	0.866	3.49
Egypt	82,999,393	1,001,449	$188,412,876,658	$2,070	0.708	3.07
Gaza Strip and West Bank	4,043,218	6,220	$4,015,865,744	$1,250	0.731	n.a.
Iran	72,903,921	1,648,195	$331,014,973,186	$4,530	0.732	1.98
Iraq	31,494,287	437,072	$65,837,434,656	$2,210	n.a.	4.00
Israel	7,441,700	20,770	$195,391,755,461	$25,790	0.872	7.48
Jordan	5,951,000	92,300	$25,092,339,119	$3,980	0.773	3.74
Kuwait	2,794,706	17,820	$148,023,721,297	$43,930	0.891	3.88
Lebanon	4,223,553	10,452	$34,528,145,455	$8,060	0.772	5,82
Oman	2,845,415	212,460	$46,114,434,702	$17,890	0.814	2.86
Qatar	1,409,423	11,437	$98,313,183,980	n.a.	0.875	3.09
Saudi Arabia	25,391,100	1,960,582	$369,178,666,667	$17,700	0.812	1.84
Syria	21,092,262	185,180	$52,176,788,110	$2,410	0.724	2.31
United Arab Emirates	4,598,600	82,880	$230,251,878,599	$26,370	0.868	2.52
Yemen	23,580,220	527,970	$26,365,156,990	$1,060	0.508	2.64
Total	291,560,271	6,215,452	$1,835,312,120,570			

Sources: * 2009 Source: http://data.worldbank.org

** 2005 Source: UNDP 2007/2008.

*** 2010 Source: The Economist Intelligence Unit 2010.

country. In Egypt, money transfers from migrant workers add significantly to the country's revenues. In fact, there is worker migration from the poorer countries to the oil exporters, particularly to Saudi Arabia, Kuwait, and the United Arab Emirates (UAE).

The Political Environment

While the economic landscape in the Middle East significantly differs between the rich oil exporting countries and the poorer and more populated states, the political landscape is uniformly dominated by non-democratic regimes. With the exception of Israel and to some extent the Lebanon, authoritarian regimes dominate the region (Ottaway 2008). The Middle East is one of the most troubled regions in the world (Goldschmidt and Davidson 2005). Conflict, often violent, is prevalent within countries (e.g. the long-lasting Israeli–Palestinian conflict), as well as between countries (e.g. the war between Iraq and Iran). The general climate of violence and insecurity in several parts of the region has prevented many countries from growing economically and developing socially. Many Middle Eastern countries are among the worst with regard to human rights standards, civil liberties and political culture (see Table 11.1). The suppression and persecution of political opposition is as prevalent as is systematic torture and intimidatory justice (El-Dawla 2009). Although hesitant political reforms have been initiated in several countries, they lacked significance in depth and breadth. Political and economic power is still highly concentrated, preventing real political opposition, freedom of opinion and the political participation of the masses (Ottaway 2008). The recent revolution in Egypt and the violent riots in Syria, the Yemen, and Bahrain indicate that fundamental political reforms seem to be impossible within the boundaries of the existing regimes. As the pressure for reform is high in many countries of this region, political unrest is likely to continue and even to increase.

The Social Environment

Economic inequality and the political situation shape the social environment in the Middle East. Apart from the small ruling class, the majority of people in many countries live close to or below the poverty line. Although economic liberalization has spurred the economic development of many countries, it has not resulted in increasing the wealth of the population. In contrast, the living conditions of the working class have worsened significantly (Beinin 2009). Unemployment rates are generally high and the lack of jobs particularly affects the young. Since the 1950s, the population of the Middle East and North Africa (MENA) has quadrupled (World Bank 2004). The strong population growth leads to changes in the demographic structure. In contrast to many Western countries, there is a strong increase in the share of the young population. In the absence of regular opportunities for advancement, corruption is a widespread phenomenon (Richards and Waterbury 2007). Educational standards have improved over the last decade but are still comparatively low, with high rates of illiteracy (e.g. around 50 per cent in Egypt). Accordingly, social security systems are poorly developed, making modern health care services rather inaccessible to the masses. Most of the social transfers are subsidy programmes for food, housing and energy (Devlin 2010). Oil-rich countries such as the UAE, Qatar and Kuwait are exceptions to the rule, in that they distribute parts of the wealth generated by oil revenues to the population.

The lack of equal opportunities, widespread corruption and political repression have fuelled ethnic and religious tension in the Middle East. Conflicts are not only prevalent between religious denominations such as the Jewish and the Muslim parts of the population in Israel but there are also many conflicts between the different Islamic groups, for example between the Shiite majority and the (ruling) Sunni minority in Bahrain. Finally, religious, social, economic and political tensions culminate in the threat of terrorism, which is widespread across the region.

The Aviation Industry in the Middle East

In line with the fast growth of the population, the aviation sector in the Middle East has grown tremendously over the past decade (Belobaba and Odoni 2009). As an exception, it even experienced a growth of air travel in the year 2009 (AACO 2010 and Boeing 2010). Forecasts predict that the annual growth rate of aviation in the Middle East will be about 6.8 per cent for the next 20 years (Airbus 2010).

In addition to population growth, investment in tourism and commerce are the main drivers of aviation growth. Especially development in the UAE and Egypt has made the Middle East the fastest growing tourist market in the world over the past decades. International tourist arrivals increased from 9.6 million in 1999 to 54.4 million in 2011 corresponding to a current market share of 5.6 per cent (UNWTO 2010). Almost all tourists arrive by plane, making long-haul flights of full service network carriers (FSNC) the most important means of transportation. FSNCs from the Gulf States, such as Emirates and Etihad, went through an unprecedented period of strong growth. Although this development has cooled down during the recent economic crisis, this trend is expected to continue in the long run. The growth in the air goes along with the development of infrastructure on the ground. Several Gulf Sates aim at diversifying their economies by investing in tourism and commerce (Vespermann, Wald and Gleich 2008). A prominent example is the Dubai World Central Al Maktoum International Airport project, which will not only result in an international airport but also in a fully integrated, intermodal logistic hub. Upon completion, this airport was originally supposed to be the largest one on earth, with a passenger capacity of 160 million people per year and about 12 million tons of capacity for air cargo. Due to the recent economic crisis, the completion of the airport had to be postponed to 2025 (Airline Business 2011). Significant airport expansion is also taking place in neighbouring Abu Dhabi. After a second runway was opened in 2008, a third terminal was inaugurated in 2009. For the future, a midfield terminal is planned, which will be located between two runways. In Qatar, the existing Doha International Airport will soon be replaced by the New Doha International Airport, triplicating the current passenger capacity. Likewise, Bahrain International Airport has launched a comprehensive expansion project, including the construction of new terminals.

The future growth in the Gulf will not only depend on tourism and commerce using the region's airport as origin and destination but the long-haul network carriers in this region are also redirecting traffic between Europe and Asia and between North America and Asia via their hubs. The favourable geographic position of their hubs allows carriers like Emirates to compete on this route against established carriers like Air France or Lufthansa. In addition to the geographic advantages, a very competitive cost structure turns Middle Eastern carriers into threatening rivals for incumbent airlines in Europe and Asia (Vespermann, Wald and Gleich 2008).

Whereas the most important carriers in the Gulf Region, Emirates, Etihad and Qatar Airways, pursue a strategy of independent growth (with the exception of codeshare agreements), several flag carriers of this region recently joined one of the three leading airline alliances. Royal Jordanian's accession to the oneworld Alliance in 2007 was followed by Egyptair joining the Star Alliance in 2008 and Saudi Arabian Airlines joining the SkyTeam Alliance in 2012 (the Lebanese Middle East Airline will join the SkyTeam later in 2012).

Traditionally, the established flag carriers of the Middle East follow the business model of a full service network carrier based on a hub and spoke network operating from a strong regional basis. Especially the Gulf-based carriers Emirates, Ethihad, and Qatar Airways are building a high level service image. According to the Skytrax airline ranking, Qatar Airways is among the seven five star airlines of the world. This category stands for the highest standards of product and service delivery quality (Skytrax 2011). Emirates and Etihad fall into the four-star category, which comprises some of the world's leading airlines such as Air France and Lufthansa. The Gulf carriers offer a high service level and target premium leisure and business travellers. Nonetheless, efficient operations and low unit costs allow these carriers to offer their services at competitive prices. Another fast growing market segment in the Middle East is individualized premium air transport services delivered by carriers like Abu Dhabi based Royal Jet or Dana Executive Jets in the emirate of Ras Al Khaimah. However, the high number of (less affluent) immigrant workers and the increasing (but less affluent) young local population recently became the focus of attention of Middle Eastern airlines.

Low Cost Carriers in the Middle East

Compared to other regions of the world, LCCs in the Middle East are a quite recent phenomenon. In 2000, virtually no LCC operated in this region. However, this segment of the airline industry is among the fastest growing in the world. Although the recent economic crisis severely hit most airlines across the planet, the Middle Eastern low cost segment continued to grow at astonishing rates (Wall, Flottau and Compart 2009). In 2010, LCCs already accounted for 6 per cent of the total seat capacity offered in this region (AACO 2010).

Regulatory Environment

The future growth of Middle Eastern LCCs is dependent on and corresponds to a rather slow but continuing process of liberalization and deregulation of the air transport sector in the region. In 2000, the Arab Civil Aviation Commission (ACAC) launched a series of step by step liberalization initiatives with the aim of creating a Common Arab Aviation Market. These initiatives have led to granting fifth freedom traffic rights between Arab states (WTTC 2004). Several countries in the region, like Oman, Kuwait and the UAE, have granted open market access to LCCs. Nonetheless, the air transport sector is still heavily regulated and barriers to market entry, such as foreign ownership restrictions, exist (AACO 2010). Many airlines and related services are still state owned or controlled by government agencies and there is an obvious gap between the declarations of intent of international organizations such as the ACAC and the Arab Air Carriers Organization (AACO) and the implementation of related policies by the different states (United Nations 2007).

Air traffic rights are not the only barriers which may hinder the future growth of LCCs in the Middle East. In several countries, travel visa restrictions prevent people from demanding more air transport services. High costs and a significant amount of time spent to obtain a visa prevent many potential customers from making impulsive short trips.

Specific Characteristics of Low Cost Carrier Services in the Middle East

Regulatory issues are only one factor Middle Eastern airline managers must deal with when operating a LCC. In contrast to their North American and European counterparts, no dedicated secondary airports exist offering a differentiated pricing for low cost travel and a budget terminal infrastructure suited to LCCs. Airports almost exclusively are state owned and operated. As a consequence, LCCs in this region usually have to cover airport and ground handling fees similar to those for FSNC (CAPA 2008). LCCs are also not able to put pressure on prices by playing airports against each other, a strategy successfully pursued by several European and American LCCs. However, first efforts are being made towards the development of a dedicated low cost infrastructure. Following the role models of Singapore Changi Airport's budget terminal and Bordeaux Airport's 'Billi' terminal, the first terminal of the Dubai World Central Al Maktoum International Airport will be focused on LCCs. Similarly, the former main terminal of Israel's Ben Gurion Airport has been converted into a budget terminal serving LCCs and charter airlines.

Another cost-cutting measure of western LCCs is to avoid complex and expensive distribution channels. Many LCCs require their customers to book online and to pay with a credit card. Although Internet penetration in the Middle East is steadily increasing, the use of credit cards is not widespread. Local customers also prefer speaking to someone when booking their flights. Therefore, LCCs in this region strongly rely on travel agents and on call centres as distribution channels. The share of online reservations of Air Arabia, the leading Middle Eastern LCC, is only 35 per cent (Airline Business 2010). Similarly, local tastes prevent carriers from abstaining from any comfort. Carriers must offer a service level higher than that in North America and Europe. Adjustments to the pure LCC business model must also be made as a consequence of religious factors. For example, free seating is not an option in the more conservative countries of the region. Seats must be pre-assigned, as women cannot be seated on an aisle seat or beside a non-related male passenger.

Customer Base

The customer base of Middle Eastern carriers is different from that of premium carriers like Ethihad or Qatar Airways. It mainly consists of the (growing) poorer parts of the population, i.e. low income freelance workers, small businesses, students, and leisure travellers. Another important customer group is migrant workers travelling from their home countries to their host countries (Airline Business 2010). In fact, labour migration has played an increasingly important role for the Middle Eastern economies in the last decades. The countries of the Gulf Cooperation Councils (GCC), i.e. the Gulf States and Saudi Arabia, constitute the largest job markets in the region. Countries of origin traditionally were other countries in the region. Due to cultural proximity, the GCC states preferred receiving workers from Arab countries like Egypt, Syria, the Yemen and Sudan. More recently, the pattern of migration has changed, leading to an increasing share of migrant workers from India, Pakistan, Bangladesh, Indonesia, Thailand and Sri Lanka. Labour migration is essential not only for the economic development of the receiving countries but also for the countries of origin. The GCC states severely lack a local workforce. Expatriate workers account for the majority of the workers and, in some countries, of the total population. In the UAE for instance, the share of foreign population was 81 per cent in 2004 (United Nations 2006). For the countries of origin, labour migration reduces unemployment rates and generates inflows of foreign capital due to the remittances of migrant workers.

Overview of Middle Eastern Low Cost Carriers

The delineation of what exactly constitutes an LCC can be difficult (see Chapter 1 of this volume) and the Middle Eastern market for this segment is highly dynamic. Therefore the list of LCCs presented in Table 11.2 must be treated with caution.

Table 11.2 Low Cost carriers in the Middle East (as of August 2011)

Carrier	IATA Code	ICAO Code	Country	Type of service	Founded	Fleet size	Destinations
Air Arabia	G9	ABY	Sharjah, UAE	Scheduled	2003	23	67
AlMasria Universal Airlines	UJ	LMU	Egypt	Scheduled	2008	2	8
Bahrain Air	*BN*	*BAB*	*Bahrain*	*Scheduled*	*2007*	*6*	*23*
Felix Airways	FO	FXX	Yemen	Scheduled	2008	4	13
FlyDubai	FZ	FDB	Dubai, UAE	Scheduled	2008	19	39
Israir Airlines	*6H*	*ISR*	*Israel*	*Scheduled/Charter*	*1996*	*7*	*13*
Jazeera Airways	J9	JZR	Kuwait	Scheduled	2005	11	22
Kang Pacific Airlines	T7	KPA	Fujairah, UAE	Scheduled	2006	1	3
Mahan Air	*W5*	*IRM*	*Iran*	*Scheduled*	*1992*	*36*	*28*
Menajet	*IM*	*MNJ*	*Lebanon*	*Charter*	*2003*	*1*	*n.a.*
Nas Air	XY	KNE	Saudi Arabia	Scheduled	2007	15	30
RAK Airways	RT	RKM	Ras al-Khaimah, UAE	Scheduled	2006	4	5
Sama	*ZS*	*SMY*	*Saudi Arabia*	*Scheduled*	*2007*	*6*	*10*
Sun d'Or International Airlines	*2U*	*ERO*	*Israel*	*Scheduled/Charter*	*1977*	*3*	*27*
Syrian Pearl	PI	SPA	Syria	Scheduled	2008	2	4

Three older carriers, Mahan Air (Iran), Israir Airlines and Sun d'Or (both Israeli), only partly qualify as LCCs, as they mainly offer charter services or exhibit several elements of an FSNC. Likewise, Bahrain Air started as a typical LCC in 2007 but recently changed its business model, adopting several elements of a network carrier. Disregarding the carriers with a somehow unfocused business model (in italics), Table 11.2 comprises nine 'pure' LCCs operating from and within the Middle East.

There are several smaller airlines with a fleet size of between two and six serving only a handful of destinations. Syrian Pearl flies to destinations within the country and offers flights from Damascus to Egypt, Saudi Arabia and Turkey. The Yemenite Felix Airways offers domestic flights and connects Sanaa to other Middle Eastern countries. The Egyptian LCC AlMasria Universial Airlines flies two Airbus A320-232s from Cairo to other Middle Eastern and several domestic destinations. Kang Pacific Airlines launched operations in 2008. The airline flies from Fujairah (UAE) to Manila (Philippines) and Dhaka (Bangladesh). It is planned to expand services to the UK and to India. Nas Air serves more than 30 domestic destinations in Saudi Arabia, India, Pakistan, Sudan, Turkey and other Middle Eastern countries. The fleet consists of 15 aircraft (Airbus A320s and Embraer E190s, E195s). The airline also specializes in religious travel to Saudi Arabia. The number of religious tourists visiting Saudi Arabia for the Hajj (the pilgrimage to Mecca during a fixed time period each year), Umrah (the pilgrimage to Mecca which must not take place within a certain timeframe), and Eid ul-Fitr (the holidays ending the fasting period of Ramadan) is steadily increasing and reached 7.8 million in 2010.

Although LCCs only recently started to conquer the Middle Eastern markets, a few carriers have already had to cease operations due to financial difficulties. For instance, Saudi Arabian Sama was founded in 2007 but had to shut down in 2010 due to large losses and lack of liquidity (Bloomberg 2010a). Similarly, RAK Airways, based in the Emirate of Ras al-Khaimah, suspended flights as a consequence of the global financial crisis in 2009. However, the airline resumed its services in 2010 with a slightly modified business model, positioning itself between an LCC and a regional carrier (Bloomberg 2010b).

Air Arabia

The first Middle Eastern LCC in a narrow sense was Air Arabia. This Sharjah based airline was founded in 2003 and currently operates a fleet of 23 Airbus A320s serving more than 67 destinations.[1] In 2010, the carrier served 5.2 million passengers with a seat load factor of 83 per cent. There are 44 aircraft orders to

---

1 Data on the past and current state of Middle Eastern low cost carrier was collected from several sources. Financial data for companies listed on the stock exchange were mainly obtained from the Infinancials (2011) database and the Marketline database (2012). Additional information was retrieved from the airline's annual reports and web pages. It

Table 11.3 Key financial indicators of Air Arabia

	Total revenues	EBIT	Net profits
2010	$566,469	$42,813	$83,266
2009	$536,966	$70,277	$123,143
2008	$562,720	$78,642	$138,849
2007	$218,839	$78,642	$76,781
2006	$203,998	$28,476	$27,542
2005	$111,965	$8,538	$8,525
2004	$49,370	-$172	-$117

Note: EBIT = earnings before interest and taxes.
Source: Infinancials database 2011.

be delivered by 2016, which will more than double the existing fleet. Currently, passengers of Air Arabia are composed of 50 per cent leisure, 30 per cent business, and 20 per cent migrant workers (*Airline Business* 2010).

Air Arabia operates mainly from two hubs, Sharjah International Airport in the UAE and Mohammed V International Airport Casablanca in Morocco. The North African Air Arabia Morocco is a joint venture between local investors and Air Arabia. The carrier flies on domestic routes and connects Morocco to destinations in North Africa and Western Europe. More recently, Air Arabia founded joint ventures with an Egyptian travel company, Air Arabia Egypt, and a Jordanian travel company, Air Arabia Jordan. The parent company mainly offers connections within the Gulf region, to other countries in the Middle East, North Africa, selected sub-Saharan countries, India, Pakistan and Europe. In addition to air transport services, Air Arabia runs a low cost hotel at Sharjah International Airport and its own flight academy (Air Arabia 2011).

The company has been listed on the Dubai Financial Market since its privatization in 2007 but the Sharjah government still holds 45 per cent of the shares. Table 11.3 provides the key financial figures of Air Arabia. The airline has been profitable since 2005 and revenues increased from US$49 million in 2004 to US$801 million in 2010. The figures for 2012 underline the continued growth of the airline. Notwithstanding the political tension in the region and high fuel prices in the region, there was a 12.8 per cent increase in passengers compared to the same period the year before.

Jazeera Airways

Jazeera Airways was the second Middle Eastern LCC. Based in Kuwait, the airline started operations in 2005. It is a public company traded on the Kuwait Stock

---

needs to be mentioned that due to the highly dynamic development of the aviation industry in the Middle East, figures on destinations, fleet size, and aircraft orders change quickly.

Table 11.4 Key financial indicators of Jazeera Airways

	Total Revenues	EBIT	Net Profits
2010	$151,648	$3,481	-$9,994
2009	$160.875	-$3,481	-$28,686
2008	$178,653	$5,360	$16,277
2007	$126,653	$5,360	$8,345
2006	$74,865	$12,006	$8,703

Source: Infinancials database 2011.

Exchange. With a fleet of 11 Airbus A320s it is currently serving 22 destinations in the Middle East and in India. In 2010, it carried 1.3 million passengers and offered services in economy class and business class. Jazeera was the first airline to offer low cost services from Dubai, where it established a second hub. In 2009, when FlyDubai started its operations as the first domestic LCC, Dubai Civil Aviation Authorities revoked the licence for Jazeera to operate from its Dubai hub. In combination with the slump of passengers during the economic crisis, this led to a significant overcapacity of the airline with a seat load factor of only 66 per cent by the end of 2009. The financial figures provided in Table 11.4 reflect the difficult situation the carrier suffered during this year.

Jazeera Airways reacted to the crisis by implementing a turnaround plan which covered measures such as dropping less profitable routes, cancelling aircraft orders and cutting the workforce by 30 per cent. To remarket its excess aircraft capacity, the airline acquired an aircraft leasing company. The network was further optimized by focusing on short-haul flights (up to two hours). As a result of these measures, Jazeera Airways returned to profitability in the second half of 2010 and was able to increase its profits in 2011 and 2012.

FlyDubai

FlyDubai is the first domestic LCC of Dubai and is fully owned by the Government of Dubai. Independent from the Emirates Group, the airline was founded in 2008 and commenced operations in 2009. Since then, it has expanded quickly, with a current fleet size of 17 Boeing B737-800s and a large order backlog of 36 aircraft. It currently serves 39 destinations in the Middle East, Russia, the Sudan, Pakistan, Afghanistan, Nepal, Bangladesh, Azerbaijan, Sri Lanka and Turkey.

Conclusion

This chapter has provided an overview on the development of low cost carriers in the Middle East. Although this business model is a quite recent phenomenon in the region, the examples of fast-growing airlines, such as Air Arabia and FlyDubai, indicate the huge market potential. Compared to their European counterparts, which today have a market share of more than 30 per cent (Koch 2010), LCC market penetration in the Middle East is still low. The capacity share of LCC accounted for only 6 per cent in 2010 (AACO 2010). The future growth of LCCs in the Middle East will be spurred by continuing population growth and economic development.

Although economic and demographic factors favour the increase of LCC activities, several context factors may limit or even reverse this trend. Further deregulation and liberalization of Middle Eastern markets are necessary preconditions for LCCs to tap the future potential for growth. Based on a simulation, Adler and Hashai (2005) showed that the deregulation of the Middle Eastern air transport industry could lead to an increase in inter-country passenger flow of 51 per cent. But also intra-country traffic has a high potential for growth. This particularly applies to countries with a comparatively large population. Saudi Arabia, for example, still has a highly regulated aviation regime, but a population of more than 25 million and a large territory make it a potentially very lucrative market for LCCs.

Another factor limiting the future growth of LCCs is the increasing fuel price. The cost advantages of LCC over FSNC result from savings in all parts of the value chain (Belobaba 2009 and Koch 2010). The price of fuel as a commodity cannot be controlled by an airline (except for short-term manipulations through fuel-hedging). Today, fuel is already the single most important cost category for many airlines (Belobaba 2009). A further increase in the fuel price may marginalize the cost advantages of LCCs and endanger the viability of the entire business model.

Finally, the political and social situation will significantly influence the future development of LCCs in the Middle East. In all parts of the world, the aviation industry proved to be particularly prone to external shocks. As discussed in the introductory sections of this chapter, the current social unrest in many countries of the region results from severe social inequality, corrupt political elites and a weak industrial basis. It is very unlikely that these issues will be resolved without fundamental changes. The revolutions in Egypt, Tunisia, and Libya, as well as the violent protests in other Middle Eastern countries, show that people are no longer willing to accept this miserable situation. Whether these developments will lead to long lasting conflicts paralysing the economy or to a new economic boom remains to be seen.

References

AACO (2010). *Annual Report 2010*. Cairo: Arab Air Carriers Organization (AACO).

Adler, N. and Hashai, N. (2005). Effect of open skies in the Middle East region. *Transportation Research Part A*, 39(10), 878–94.

AEA (2006). Blueprint Middle East, AEA (Association of European Airlines). *Market Research Quarterly*, 3.

Air Arabia (2011). *Group*. Available at: http://www.airarabia.com/crp_1/group&stitle=group&pid=125 [accessed: 11 November 2011].

Airbus (2010). *Global Market Forecast 2010–2029*. Toulouse: Airbus 2010.

Airline Business (2010). Interview Air Arabia chief executive Adel Ali. *Airline Business*, 21 April 2010.

Airline Business (2011). Dubai plots 2025 Emirates hub shift. *Airline Business*, 27(8), 16.

Beinin, J. (2009). Workers' struggles under 'socialism' and neoliberalism, in El-Mhadi, R. and Marfleet, P. (eds) *Egypt: The Moment of Change*. Cairo: Zed Books, 68–86.

Belobaba, P. (2009). Airline operating costs and measures of productivity, in Belobaba, P., Odoni, A. and Barnhart, C. (eds) *The Global Airline Industry*. Chichester, UK: John Wiley, 113–51.

Belobaba, P. and Odoni, A. (2009). Introduction and overview, in Belobaba, P., Odoni, A. and Barnhart, C. (eds), *The Global Airline Industry*. Chichester, UK: John Wiley, 1–17.

Bloomberg (2010a). *Saudi Low-Cost Airline Sama to Cease Operations After $266 Million Loss*. Available at: http://www.bloomberg.com/nezs/print/2010-09-26/ [accessed: 11 November 2011].

Bloomberg (2010b). *RAK Airways to Restart Operation on October 10*. Available at: http://www.bloomberg.com/news/print/2010-08-22/ [accessed: 11 November 2011].

Boeing (2010). *Current Market Outlook 2010–2029*. Seattle, WA: The Boeing Corporation.

CAPA (2008). *Low Cost Airport Terminals Report*. Sydney, Australia: Centre for Asia Pacific Aviation (CAPA).

Devlin, J.C. (2010). *Challenges of Economic Development in the Middle East*. Singapore: World Scientific.

El-Dawla, A. (2009). Torture: A state policy, in El-Mhadi, R. and Marfleet, P. (eds) *Egypt: The Moment of Change*. Cairo: Zed Books, 120–37.

Francis, G., Humphreys, I., Ison, S. and Aicken, M. (2006). Where next for low cost airlines? A spatial and temporal comparative study. *Journal of Transport Geography*, 14(2), 83–94.

Franke, M. (2004). Competition between network carriers and low-cost carriers – retreat, battle, or breakthrough to a new level of efficiency? *Journal of Air Transport Management*, 10(1), 15–21.

Goldschmidt, A. and Davidson, L. (2005). *A Concise History of the Middle East*. Boulder, CO: Westview Press.

Infinancials Database (2011). Starting page. Available at: http://www.infinancials.com [accessed: 11 November 2011].

Koch, A. (2010). Aviation strategy and business models, in Wald, A., Fay, C. and Gleich, R. (eds), *Introduction to Aviation Management*. Berlin: Lit Verlag Dr. W. Hopf, 145–84.

Marketline (2012). Company profiles. Available at: http://www.marketline.com [accessed 25 March 2013].

Mason, K.J. and Morrison, W.G. (2008). Towards a means of consistently comparing airline business models with an application to the 'low cost' airline sector. *Research in Transportation Economics*, 24(1), 5–84.

Ottaway, M. (2008). Evaluating Middle East reform: Significant or cosmetic, in Ottaway, M. and Choucair-Vizoso J. (eds), *Beyond the Façade: Political Reform in the Arab World*. Washington, DC: Carnegie Endowment for International Peace, 1–15.

Pitfield, D.E. (2008). Some insights into competition between low-cost airlines. *Research in Transportation Economics*, 24(1), 5–14.

Porter, M.E. (1980). *Competitive Strategy, Techniques for Analyzing Industries and Competitors*. New York: Free Press.

Richards, A. and Waterbury, J. (2007). *A political economy of the Middle East*. Boulder, CO: Westview Press.

Skytrax (2011). The world's largest Airline and Airport review site. Available at: http://www.airlinequality.com [accessed: 14 November 2011].

United Nations (2006). *United Nations Expert Group Meeting on International Migration and Development in the Arab Region*, UN/POP/EGM/2006/02, Beirut.

United Nations (2007). *ESCWA Sudy on Air Transport in the Arab World*. New York: United Nations Economic and Social Commission for Western Asia (ESCWA).

UNWTO (2010). *UNWTO Tourism Highlights* (2010 Edition). Madrid: World Tourism Organisation.

Vespermann, J., Wald, A. and Gleich, R. (2008). Aviation growth in the Middle East – impacts on incumbent players and potential strategic reactions. *Journal of Transport Geography*, 16(6), 388–94.

Wall, R., Flottau, J. and Compart, A. (2009). Air transport growth sector: Middle East low-fare airlines. *Aviation Week and Space Technology*, 171, 17.

Wensveen, J.G. and Leick, R. (2009). The long-haul low cost carrier: A unique business model. *Journal of Air Transport Management*, 15(3), 127–33.

World Bank (2004). *Unlocking the Employment Potential in the Middle East and North Africa: Toward a New Social Contract*. Washington, DC: The World Bank.

WTTC (World Travel & Tourism Council) (2004). *Aviation in Egypt: The Impact on Travel & Tourism, Jobs and the Economy*. London: WTTC.

PART VI
Conclusion

Chapter 12

Conclusions: The Future of Low Cost Carriers

Markus Landvogt, Sven Gross and Michael Lück

Without doubt, and as all the previous chapters of this volume highlight, low cost carriers have truly become a phenomenon all over the world despite showing great variations in their business models and being at different stages in their life cycles subject to economic prosperity and regulations in the aviation sector.

This final chapter draws conclusions from the successful introduction of low cost carriers (LCCs) in the world, outlines the consequences for the aviation industry and provides an outlook into the future of LCCs in the aviation industry.

The Introduction of Low Cost Carriers

Since the early start of LCCs in the US there has been considerable interest from a wide range of stakeholders such as consumers, travellers, the aviation industry, governments and destination management organizations around the world, not to mention media, environmental organizations and investors, all of whom have been fascinated by the emergence of LCCs. Interested stakeholders who have not yet seen the LCC development close to their doorstep look with envy to regions where LCCs are established as an attractive alternative form of travel.

From a demand perspective, there are two groups of travellers. In one group people need to travel to a certain destination and are looking for suitable means of transport. The other group comprises those whose destination choices are primarily influenced by availability and provision of affordable transportation options to a set of destinations.

The LCCs' success coincides with attracting travellers from the latter group, travellers who are interested in affordable air fares to destinations previously not offered, some of whom are first-time air travellers, who would not have flown otherwise (Franke 2004). From the beginning, LCCs have been targeting the segment of travellers attracted by three features: air transportation: accessibility of destinations outside travellers' usual environment; and low fares.

The literature tends to agree that a general definition of LCCs is inevitably rather ambiguous due to the fact that there is not one clear and distinct business model of an LCC, but instead a range of product and business differentiations when comparing LCCs. What all LCCs have in common is a very strong focus on managing the costs of all their operations at a low – if not the lowest possible – level. Based on this assumption, different LCCs have pursued their business models over time by aligning them in slightly different ways to the key principles of the low cost carrier model (see, in particular, Chapter 1).

Obeying these key principles, the business models of LCCs focus on the reduction of operating costs and maximizing revenue generation. As a consequence, LCCs have obtained a leadership role with regard to cost and labour productivity, which is what allows them to offer air travel at a much lower cost. Therefore, low fares have become paramount and have overtaken comfort; service and punctuality as the primary decision-driver for at least short-haul travel (Thanasupsin et al. 2010).

Obviously, some of the key principles are impacting each other and cannot be seen in isolation from each other. Given such a long list of principles, it can be imagined that, by slightly or totally modifying one of the characteristics, a different business model emerges. For most of the principles, a LCC can be found which does not comply, or complies only slightly, with just that principle.

Even when looking at those airlines which are considered pure LCCs, one would find that some of the key principles are softened, changed or neglected (O'Neil-Dunne et al. 2008). Thus, it can be concluded that no two LCC business models are the same.

For example, Southwest Airlines, WestJet, Ryanair, easyJet, Buzz Air, AirAsia, and Tiger Airways are seen as true LCCs. However, Ryanair's business model is regarded as the purist amongs the LCC models (O'Neil-Dunne et al. 2008). Other low cost airlines have been able to find and create their own niche market with a slightly different business model.

Some airlines, including LCCs, are trying to reposition themselves in the wide continuum from a pure LCC model to a classic legacy airline. These airlines have adapted some of the key principles of LCCs but at the same time offer additional services which are typical for full service airlines (FSAs).

This highlights the difficulties inherent in labelling an airline business model 'low cost carrier'. Mason and Morrison (2008) suggest, therefore, a benchmark metrics based on a set of measurable indices to analyse and compare the LCCs' business models.

The Implications of Low Cost Carriers

The LCCs have brought some significant implications to the aviation markets. Fundamental for the development of LCCs were the easing of the regulatory environments and the LCCs' pushing the boundaries for the liberalization of the regulatory frameworks at the same time. Deregulation is seen as a catalyst for the introduction of LCCs, although deregulation is not a necessary condition for their success.

The earlier the market liberalization happened, the further the segment of LCCs developed. Clearly, the US has been at the forefront of deregulation, followed by the European Union (EU). With cross-border joint ventures, LCCs try to overcome and push the boundaries of restrictions in Southeast Asia, the Middle East and South America (Knibb 2008). It is only a question of time until regulations are further eased, for example in China, India and Africa.

The continuous growth of LCCs in terms of routes and airports serviced by offering flights at lower costs has cut into the market share of FSAs. LCC routes became an attractive and cheaper alternative for travellers, including business travellers, who otherwise had no other choice than to fly with a FSA.

In fact, FSAs have developed a range of responses to the challenges from LCCs. First of all, the expansion of LCCs has forced full service carriers (FSCs) to reassess their own operational costs and yields. Doganis (2001: 222) suggests that 'cost reduction is no longer a short-term response to declining yield or falling load factors. It is a continued and permanent requirement if airlines are to be profitable'. FSAs lowered their costs, or the number of their full-time employees, to better compete with LCCs. Aer Lingus is seen as an example where a FSA adopted an LCC model and practically became an LCC as a response to arch rival Ryanair (O'Neil-Dunne et al. 2008).

FSAs benefit from their hub-and-spoke systems providing connections between any two points of the network. The ability to bundle demand from several places of origin provides competitive advantages for these network carriers (Franke 2004). Most recent consolidations in the US and Europe in the form of integrations or mergers suggest that there is a reliance on economies of scale for hub-and-spoke business models. Although the focus of the network carriers remains on the valuable network customer, they are also promoting spare capacities on a point-to-point basis at competitive prices.

Some FSAs choose to establish their own LCC subsidiaries to meet the challenges from low cost competitors. Graham and Vowles (2006) analysed the phenomenon of low cost 'carriers within carriers' (CWCs) and the evolution of their strategies. A CWC offers the advantage of a subsidiary with significant lower labour costs and provides a direct competitive response to LCCs already operating or planning to operate in the market. However, there has been only limited evidence that low cost CWCs constitute an effective market response to low cost competition. A few of those low cost CWCs which emerged in the US and Europe were withdrawn from the market after some years (see Chapter 1 of this volume). The more recent introduction of low cost CWCs into Australasia seems to be more successful, probably due to high economic growth rates combined with a backlog of demand for travel.

The LCCs brought innovations to the aviation market by leveraging one-way fares, user-friendly online booking facilities and electronic tickets and boarding passes as a common feature of air travel.

The preference of some LCCs for secondary airports has helped to mitigate the pressure on overcrowded airports. Potential new entrants, including LCCs, find it harder in congested airports to obtain a sufficient number of attractive slots. Secondary airports not only provide better access, fewer delays and fast turnaround times but they also tend to charge airlines less for their services. Some airports have even attracted LCCs by constructing dedicated – so-called 'low cost' – terminals.

LCCs at secondary airports contribute to the economic development of the region in the vicinity of these airports. Beyond direct and indirect effects on employment, induced effects arise from business decisions caused by new air transport links.

Due to LCCs' focus on efficiency, direct travel and costs, they have played an important role in exploiting travel affluence and maintaining growth in otherwise mature markets by driving down travel costs (Harbison and McDermott 2009). Also Franke (2004) sees evidence for the impression that lower fares offered by LCCs, as well as respective reactions by network airlines, have been stimulating demand. The LCCs' principal focus on costs has contributed to passenger air transport remaining affordable despite rising oil prices.

The business models of the LCCs and the competition they brought to the aviation market have forced the aviation industry to focus more on innovation, a development that might not have occurred otherwise. Innovation efforts in the whole aviation industry have been increased by the pressure caused by LCCs (Franke 2004, 2007).

The Future of Low Cost Carriers

Although LCCs are facing some challenging opportunities and threats in the near future, the aviation experts do not differ a lot when summarizing LCC development so far:

- It is beyond dispute that low cost methodology has transformed the aviation sector. The expansion of the seat capacity in the sector has been accounted for by LCCs. The growth of the low cost market increased the intensity of competition for the short-haul, low price passenger (Harbison and McDermott 2009).
- The LCC phenomenon clearly has a critical mass and will continue to grow .... It is equally clear that air travel is a commodity. Low prices are the primary decision-driver for travel (O'Neil-Dunne et al. 2008).
- LCCs have exploited travel affluence and played an important role in maintaining growth in otherwise mature markets, boosting air travel in the North American, West European and Australasian markets, where they have played a key role in driving down travel costs (Harbison and McDermott 2009).
- Economic growth in South East Asia, China, India, the Middle East, East Europe and latterly Africa and South America has reached record highs

> over the past decade. Given that average incomes, while growing, still remain relatively low in most parts of these regions, LCCs have tapped into a latent pool of demand and enabled their citizens to take to the skies earlier than would have been the case under a more traditional business model (Harbison and McDermott 2009).

The success of the low cost model seems likely to continue. There is a wide agreement that the LCC market has the potential to grow further. In the future, however, the LCC sector will face various challenges, among them the following.

Safety

For LCCs as well as for legacy airlines all aspects of safety are paramount. The low cost label might entice consumers to infer that cost savings could undermine compliance with required safety procedures. However, this is monitored by government agencies; and since extensive media coverage appears to be one of the LCCs' main communication strategies, a negative coverage of safety records would have more damaging implications. In 2011, for example, all the aircraft of Tiger Airways Australia were grounded due to safety concerns.

The Cost of Fuel

One of the major risk factors for the airline industry and in particular the LCCs is the cost of crude oil. According to the International Air Transport Association (IATA), in 2011, the sharp rise in oil prices was one of the main factors that slashed the airline industry's profit expectations that year. Jet fuel, the primary fuel used in commercial aviation, is more highly priced than crude oil, but the two are normally highly correlated.

Oil prices over the past few years have shown two predominant developments. Firstly they are generally upward-trending. Secondly the past few years have seen considerable volatility in oil prices. Both trends, individually and together, pose a major threat to the airline sector, not exempting LCCs.

As far as fuel consumption of an aircraft type is concerned, the business model of an airline has hardly any, or only marginal, impact (Doganis 2001 and Gross and Schröder 2008). The cost of fuel is mainly determined by the type of aircraft and the efficiency of the engines. Due to their overall lower operating costs, the share of fuel costs of their total costs is proportionally higher for LCCs than for other airlines. Since fuel is the principal single operating expense, LCCs are more vulnerable to volatile and rising fuel prices than FSAs (Harbison and McDermott 2009). Hedging fuel prices is a viable option for managing this risk in the short and medium term.

Although LCCs may be able to compensate for some of the risk with a higher seat load factor or with a denser seat configuration than FSCs, eventually and inevitably, some of the rising fuel costs will need to be charged back to travellers with higher ticket prices, or in the form of fuel surcharges. This could lead into a vicious cycle by slackening demand, because LCCs are targeting market segments characterized by higher price sensitivities.

The future development of the LCC sector depends on how well this risk is managed. It is also clear that in order to lessen this risk, a substantial yield must be generated in other parts of the business, for example with ancillary revenues.

Environmental Impacts

Although airline chief executives seem to be less concerned with costs and charges related to the environmental impacts of air transportation, the environmental issues are likely to have a much greater impact on the shape of the LCC market in the future (Harbison and McDermott 2009).

With regard to pollution and noise emissions LCCs are quite good at meeting national standards as long as they prefer deploying newly manufactured aircraft and reject fuel-thirsty older aircraft. Newer aircraft engines are usually more fuel efficient and less noisy. However, there are indications that sooner or later airlines will be made accountable for their greenhouse gas emissions and that they will be unable to prevent this from happening. So far, the emissions from international aviation have been excluded from the Kyoto Protocol agreement, and at the Copenhagen conference a worldwide agreement could not be reached.

On a regional or continental level, emission-trading schemes aim at incorporating aviation greenhouse gas emissions. For instance, the EU aviation emissions are included in the EU Emissions Trading Scheme (ETS), effective 1 January 2012 (European Union 2008). This will require airlines to monitor and report their emissions, and purchase and potentially trade some of the emissions allowances. In fact, the EU ETS is a first step towards charging airlines for carbon emissions and providing incentives for the airlines to lower their emissions. The EU ETS makes costs of carbon emissions proportional to the fuel consumption per passenger kilometre, adding another factor to the cost of fuel.

Government Charges and Fees

Without doubt the aviation industry delivers immense social and economic benefit. Economies would not perform in the same way without the aviation sector. The fact that air passenger services have become a common good makes the sector also enticing for governments to search for other sources of income. The air passenger duty introduced in the United Kingdom in 2007 and the air passenger tax (*Luftverkehrsabgabe*) in Germany, which came into effect in January 2011, are two examples for charges laid on passengers for the purpose of generating additional revenue for the governments. Such charges disproportionately burden passengers of LCCs. As a consequence of the German *Luftverkehrsabgabe*, Ryanair ceased domestic flights in Germany.

Ancillary Revenues

As margins from the sale of seats keep shrinking, ancillary revenues have become a main focus for LCCs, but also for FSAs. After detaching the services from the main product (the mere flight) the ancillary revenues are generated by offering extra services to passengers who are prepared to pay extra for these services (see Chapter 1).

These extra services can comprise flight-related services such as excess baggage, in-flight food and beverage, in-flight entertainment or Internet access during the flight. At the same time airlines offer services pre-flight (e.g. travel insurance, credit cards or iPhone and smartphone apps) or post-flight (e.g. accommodation, car rental, transfers, tickets), allowing the airlines to vertically diversify in the tourism sector.

Although many airlines are focusing on generating ancillary revenues, the purists in the LCC sector seem to be in a leading position. Clearly, the most simplified and unbundled flight product has many more options for generating ancillary revenues than services that have been inclusive but are suddenly offered separately. Thus, the purer the LCC business model the easier it is to implement the provision of extra services through profit centres. Ryanair, for example, is already achieving 22 per cent of its total revenues through ancillary revenues (Harbison and McDermott 2009). These revenues are the bread and butter business for LCCs, but FSAs are increasingly implementing such additional revenue streams.

Competition Within the Sector

In most regions the development of LCCs occurred in stages. At the initial stage, LCCs primarily dipped into new traveller markets before competing with legacy airlines and drawing demand from their markets. With the continuous growth of LCCs in terms of routes and frequencies, some markets are becoming increasingly saturated, especially in North America and Europe.

As Binggeli and Pompeo (2002) show LCCs, like FSCs, prefer monopoly situations on their routes. Since LCCs can gain advantage from direct competition with one FSC, such routes are the second most preferred option. LCCs try to avoid direct competition with other LCCs on the same route. There have been examples where direct competition between LCCs has caused one airline to withdraw as seen with Pacific Blue on domestic routes in New Zealand in 2010 after Jetstar introduced domestic flights. Expansion strategies in markets where consumers are already well served with routes are risky and therefore not the best survival strategy.

Survival Strategies of Low Cost Carriers

There appear to be some significant advantages for LCCs which have been first entrants in particular markets, making it more difficult for new entrants. When it comes to benefiting from being in a monopoly supplier position on at least some routes or in some markets, LCCs are in no way different to any legacy airline. It seems that this is still an important factor to make a business viable.

Economies of scale provide advantages for LCCs. Some of these airlines recently placed huge aircrafts orders, for example AirAsia's 2011 order of 300 A320 aircraft (CAPA 2011) or Lion Air's order of 230 B737 aircraft (Airports International 2012). Economies of scale place larger companies in a much better position with regard to managing the risks across the route network than a small airline relying on a few aircraft and routes. An aircraft deployed on a route that suddenly becomes less lucrative due to competition or changes in other external factors, e.g. airport fees or air passenger duties, can more easily be moved to markets which promise to be more lucrative.

Potential for Low Cost Carriers

There still seems to be a significant growth potential in the LCC market, although the views on its size vary widely. Up to 70 per cent of the matured continental markets in Europe and the US could be served by LCCs (Franke 2004). A Delphi panel on future trends in the EU forecasts that more than half of the intra-EU traffic will be served by LCCs by 2015 (Mason and Alamdari 2007). This seems to be in line with previous forecasts saying that the total share of low fares traffic in Europe would reach over 40 per cent by 2010 (ELFAA 2004). These predictions leave the European LCC sector with a potential market share of between 40 per cent and 70 per cent of the short-haul markets. Assuming that some of the business is indeed picked up by special offers from FSAs, it still provides a sizeable potential for the LCCs.

These factors underline the fact that stronger LCC growth is expected to emerge in areas of the world currently still underserved by airlines.

Long-haul Low-cost

Although the idea was initially rejected by industry and academics, new attempts at implementing low cost strategies in long-haul markets are being undertaken (Gross and Schröder 2008). After some unsuccessful ventures early in the 1980s, for example by Laker Airways Skytrain on transatlantic flights, and in the last decade with Oasis Hong Kong Airlines, EOS and Maxjet, long-haul low-cost is increasingly receiving attention.

Clearly, some of the key LCC principles are generally irreconcilable with long-haul operations. Fast turnaround times to achieve higher aircraft utilization during the day are less relevant for long-haul flights as many sectors are overnight flights anyway. More efficient narrowbody aircraft such as the new Boeing 737-900ER, which can cater for transatlantic flights, enable LCCs to look beyond their current route structures (O'Neil-Dunne et al. 2008). Airbus A330 and next generation long-haul aircraft Airbus A350 und Boeing B787 are playing an important role in the plans for long-haul low cost operations (Harbison and McDermott 2009). Secondary airports and the lack of interlining and network connections limit the potential customer base for long-haul low cost flights.

In particular in the Asia Pacific region airlines are entering into the long-haul low cost market. Jetstar, a Qantas subsidiary, codeshares and interlines with Qantas. Air Asia X, operating as a franchise of Air Asia, relies on passenger feed from Air Asia short-haul regional business. Scoot, the long-haul low cost airline launched by Singapore Airlines in November 2011, seems to follow the strategy of product diversification and competitive responsiveness. For Gross and Schröder (2008) it is unlikely that the low cost business model on long-haul routes will achieve similar market shares as seen for LCC in continental traffic. But obviously these operations are successfully tapping into a market niche with their long-haul flights.

Final Outlook

Considering the future of LCCs in the aviation sector provides some interesting perspectives. The LCC business models will see further adjustments in order to find the best fit for the survival of individual companies in their markets. Competition with FSCs and within the LCC sector in the mature markets will help to keep the sector innovative. LCCs in developing markets will contribute to the airline sector and therefore also to the economic development in those regions. Long-haul low cost carriers will find their niche and add to the pressure on full service network carriers.

The upward trending and volatile costs of fuel will remain a significant risk factor for the airline sector, but at the same time force airlines to become as cost effective as they possibly can.

Innovative changes for the LCC business may occur with lower cost competitors targeting the duopoly of the aircraft manufacturers. In June 2011 Michael O'Leary announced that Ryanair could order as many as 400 aircraft from Chinese manufacturer Comac if the company can design a 200-seater variant of a new jet acceptable to the LCC customer.

References

Airports International (2012). Lion Air orders 230 aircraft. Stamford, UK: Key Publishing. Available at: http://www.airportsinternational.com/2012/02/lion-air-orders-230-aircraft/ [accessed: 15 February 2012]

Binggeli, U. and Pompeo, L. (2002). Hyped hopes for Europe's low-cost airlines, in *McKinsey & Company Quarterly Report*, November 2002. Available at: https://www.mckinseyquarterly.com/Transportation/Sectors/Hyped_hopes_for_Europes_low-cost_airlines_1231 [accessed: 3 December 2011].

CAPA (2011). Paris 2011: AirAsia eyes fleet of over 300 aircraft by 2020 with new A320neo. Sydney, Australia: Centre for Asia Pacific Aviation (CAPA). Available at: http://www.centreforaviation.com/analysis/paris-2011-a320neo-order-to-ensure-airasia-has-fleet-of-over-300-aircraft-by-2020-53679 [accessed: 15 February 2012].

Doganis, R. (2001). *The Airline Business in the 21st Century*. London: Routledge.

ELFAA (2004). *Liberalisation of European Air Transport: The Benefits of Low Fares Airlines to Consumers, Airports, Regions and the Environment*. Brussels: European Low Fares Airlines Association. Available at: http://www.elfaa.com/documents/ELFAABenefitsofLFAs2004.pdf [accessed: 12 May 2011].

European Union (2008). *Directive 2008/101/EC, European Official Journal of the Union L8/3*. Available at: http://eur-lex.europa.eu/LexUriServ/LexUriServ.do?uri=OJ:L:2009:008:0003:0021:EN:PDF [accessed: 14 May 2011].

Franke, M. (2004). Competition between network carriers and low-cost carriers – retreat battle or breakthrough to a new level of efficiency? *Journal of Air Transport Management*, 10, 15–21.

Franke, M. (2007). Innovation: The winning formula to regain profitability in aviation? *Journal of Air Transport Management*, 13, 23–30.

Graham, B. and Vowles, T.M. (2006). Carriers within carriers: A strategic response to low-cost airline competition. *Transport Reviews*, 26(1), 105–26.

Gross, S. and Schröder, A. (2008). Low cost business model on long-haul routes – a promising market segment?, in *Proceedings of the 12th Annual World Conference Air Transport Research Society (ATRS)*, Athens, Greece.

Harbison, P. and McDermott, P. (eds) (2009). *Global LCC Outlook Report*. Sydney, Australia: Centre for Asia Pacific Aviation (CAPA).

IATA (2011). Forecast profits revised down sharply, IATA Economics, Financial Forecast June. Geneva: International Air Transport Association. Available at: http://www.iata.org/whatwedo/Documents/economics/Industry-Outlook-June2011.pdf [accessed: 14 May 2011].

Knibb, D. (2008). Border crossing. *Airline Business*, 24(5), 58–60.

Mason, K.J. and Alamdari, F. (2007). EU network carriers, low cost carriers and consumer behaviour: A Delphi study of future trends. *Journal of Air Transport Management*, 13, 299–310.

Mason, K.J. and Morrison, W.G. (2008). Towards a means of consistently comparing airline business models with an application to the 'low cost' airline sector. *Research in Transportation Economics*, 24(1), 75–84.

O'Neil-Dunne, T., O'Toole, K. and Schonland, A. (eds) (2008). *The Future of Low Cost Carriers*. Sutton, UK: FlightGlobal.

Thanasupsin, K., Chaichana, S. and Pliankarom, S. (2010). Factors influencing mode selections of low-cost carriers and a full-service airline in Thailand. *Transportation Journal*, 49, 35–47.

Appendix 1.1

List of Currently Operating Low Cost Airlines (109)

1Time
Aer Lingus
Air Arabia
Air Arabia Egypt
Air Arabia Maroc
AirAsia
AirAsia X
airBaltic
Air Berlin
Air Busan
Air Do
Air India Express
Air Italy
Air Japan
Air One
Aires
Airphil Express
AirTran
Allegiant Air
Anadolu Jet
Avianova
Azul Lineas Brazilian
Belle Air
Blue Air
Blu Express
bmibaby
Cebu Pacific Air
China West Air
Citilink
Corendon Airlines
Dana Air
Eastar Jet
Easyfly
easyJet
easyJet Switzerland
Felix Airways
Fly540
Flybe
Flydubai
Frontier Airlines
Fuji Dream Airlines
Germanwings
GoAir
GOL aéreas inteligentes
Helvetic Airways
Iceland Express
IndiGo
Indonesia AirAsia
Interjet
Intersky
JAL Express
Jazeera Airways
Jet2.com
Jet4you
JetBlue
JetKonnect
JetLite
Jetstar Airways
Jetstar Asia Airways
Jetstar Pacific
Jin Air
Kingfisher Red
kulula
Lion Air
Mango Airlines
Mihin Lanka
Monarch Airlines
Nas Air
Niki
Nok Air
Norwegian
Onur Air
Pacific Blue
Pegasus Airlines
Polynesian Blue
Ryanair
Sky Express
Skymark Airlines
Skynet Asia Airways
Smart Wings
Southwest Airlines
SpiceJet
Spirit Airlines
Spirit of Manila Airlines
Spring Airlines
Star Flyer
Sun Country Airlines
Sunwings Airlines
sverigeflyg.se
Thai AirAsia
Thomson Airways
Tiger Airways
Tiger Airways Australia
Transavia Airlines
Transavia France
USA3000 Airlines
Valuair
Virgin America
Virgin Australia
Viva Aerobus
Volaris
Vueling Airlines
Webjet Linhas Aéreas
WestJet
Wind Jet
Wizz Air
Wizz Air Bulgaria
Wizz Air Ukraine
Zestair

Appendix 1.2

List of Former Low Cost Airlines

(including start ups/never started)

Adam Air
Aeris
AerOasis
Air Andalucia
Air Deccan
Air Luxor Light
Air Madrid
Air Next
Air One Feeder Airlines
Air Polonia
Aladia Airlines
Alma de México
America West
ATA Airlines
Atlas Blue
Aviacsa de Mexico
Avolar
BackpackersXpress
Bahrain Air
BasiqAir
BerlinJet
Blue Wings
Bra Transportes Aereos
Buzz
Centralwings
CityJet
Civair
Clickair
Compass Airlines
Continental Lite
Dauair
DBA
Delta Express
Duo
Eos Airline
EU Jet
Feel Air
Fly Eco
Fly Me
Fly Nordic
Flyglobespan
Flying Finn
Freedom Air
Germania Express
GetJet Poland
Go
Goodjet
Greyhound Air
Hellas Jet
HLX
Hop
mpulse Airlines
Independence Air
Jetgreen
JetsGo
K2000
Kiwi Travel International Airlines
Kiwijet
LAN Express
Maersk Air
Mandala Airlines
Maxjet
Menajet
MetroJet
Mexicana Click
Mexicanalink
MyAir
Now Airlines
Oasis
Pal Express
Red1 Express
Sama Airlines
Shuttle by United
Silverjet
Skybus
Skyeurope Airlines
Smartjet Airways
Snowflake
Song Air
Southeast Airlines
Spirit of Balkan
Sterling Airlines
Tango
Ted
Thomsonfly
Transavia Denmark
Transgulf Express
V Bird
Virgin Express
VistaJet
Viva Macao Airlines
VolareWeb
Zip
Zoom Airlines

Index